Pirate Politics

The Information Society Series

Laura DeNardis and Michael Zimmer, series editors

Interfaces on Trial 2.0, Jonathan Band and Masanobu Katoh

Opening Standards: The Global Politics of Interoperability, Laura DeNardis, editor

The Reputation Society: How Online Opinions Are Reshaping the Offline World, Hassan Masum and Mark Tovey, editors

The Digital Rights Movement: The Role of Technology in Subverting Digital Copyright, Hector Postigo

Technologies of Choice? The Capabilities Approach and Information and Communication Technologies for Development (ICT4D), Dorothea Kleine

Pirate Politics: The New Information Policy Contests, Patrick Burkart

Pirate Politics

The New Information Policy Contests

Patrick Burkart

The MIT Press
Cambridge, Massachusetts
London, England

MIT Press books may be purchased at special quantity discounts for business or sales promotional use. For information, please email special_sales@mitpress.mit.edu.

This book was set in Stone Sans and Stone Serif by the MIT Press. Printed and bound in the United States of America.

Library of Congress Cataloging-in-Publication Data

Burkart, Patrick, 1969–
Pirate politics : the new information policy contests / Patrick Burkart.
    p.   cm — (The information society series)
Includes bibliographical references and index.
ISBN 978-0-262-02694-9 (hardcover : alk. paper)   1. Piratpartiet.   2. Piratenpartei Deutschland.   3. Internet—Political aspects—Sweden.   4. Internet—Political aspects—Germany.   5. Information society—Political aspects—Sweden.   6. Information society—Political aspects—Germany.   7. Political parties—Sweden.   8. Political parties—Germany.   9. Sweden—Politics and government—21st century.   10. Germany—Politics and government—21st century. I. Title.
JN7995.P57B87 2014
324.243'08—dc23
2013027450

10 9 8 7 6 5 4 3 2 1

For David

Contents

Preface

The idea for this book was planted in 2009 as I was completing the manuscript for *Music and Cyberliberties* and entertaining the notion of an international umbrella movement for cyberliberties. As that book was going to press, news of the upstart Swedish Pirate Party began hitting the media, first on tech blogs and later in the mainstream press. As news of the group propagated and its influence spread to the European Parliament, Sweden and other eurozone countries faced new and recurring threats to cyberliberties, further stoking the Pirates' grievances and increasing their salience. Had the umbrella movement arrived so quickly? *Pirate Politics* had to be written to find out.

So I spent the 2012 school year researching pirate politics and teaching and living in Sweden while writing this book. To some of my students, the Pirate Party is a joke or a prank—of casual interest, but not to be taken seriously. The party's single issue of digital rights flouts their expectation that politics should be more broadly engaged with society's other problems, especially the economic crisis gripping the eurozone. However, those students who are members of the Pirate Party take their memberships *very* seriously. In class, our lively, well-informed debates and discussions about cyberliberties were consequential for the students, most of whom lived in Sweden, Germany, or another country with a budding pirate party. We debated pirate politics in real time in response to the Anti-Counterfeiting Trade Agreement (a new, plurilateral trade agreement serving as a Trojan horse for copyright reforms), new data-retention and surveillance laws, and other contemporary cyberliberties challenges.

In the closing months of 2012, the Swedish Pirate Party was struggling to rekindle its initial spark, while the German Pirates risked their newfound popularity by sharing their internal squabbles and disagreements publicly on online discussion boards. The mature pirate parties appear to be at risk of breaking down, while new ones in the Czech Republic and elsewhere are in ascendancy. Popular commentators will, undoubtedly, continue to marvel that anarchistic or apolitical geeks found a political footing to begin with, and will look no further than the flash in the pan.

Determining whether the Pirates are like an evanescent wave that has already decayed, or will persist in one form or another, is not the point of this book. It is also not my intention to rally supporters, defend detractors, or make predictions. Instead, I wish to explore the international contexts and contents of resistant forms of political and cultural agency aligned against structures and processes of media and communication perceived to be toxic and unjust. Looking for, finding, and describing the persistent features of agency in the flux—the recurring assertions of autonomy and solidarity—are the real objectives here.

Another objective is to rehabilitate prior discourses of political and social theory concerning the nature of "new politics," especially sociological theories of social movements, and to try them on for size once more. This goal will grate against most empirical approaches to political systems, and against game-theoretical and neorealist perspectives on international relations. I make no special claims to expertise in those fields. But I ask for no forbearance either, and invite all appropriate critiques if they can help us better understand what we agree that we are observing.

Acknowledgments

I thank my colleagues and students at Karlstads Universitet Medie-Och-Kommunikationsvetenskap for providing a departmental home while I researched and wrote this book. Their support and kindness was invaluable. Göran Bolin, Christian Christiansen, and Miyase Christensen, Leif Dahlberg, Peter Jakobsson, Mats Nilsson, Klas Sandell, Martin Fredriksson, Stefan Larsson, Ad van Loon, and Joep van der Veer provided essential feedback and guidance for the duration of the project. Steve Bales offered excellent library support. Jonas Andersson blazed a research trail in Sweden that I tried to follow. He generously took every opportunity to offer insights and corrections to my drafts. Stephan Urbach, Marina Weisband, Julia Reda, Carsten Koschmieder, Eva Schweitzer, and Joel Schalit gave me important access to German perspectives on pirate politics.

The Melbern G. Glasscock Center for Humanities Research, at Texas A&M University, provided support for this project. I benefited from early opportunities to present my ideas at the International Communication Association—Popular Communication Division and the Global Fusion conferences, and also at the University of San Diego Department of Communication Studies, the

University of Montana Department of Communication Studies, Northwestern University in Qatar's Department of Journalism and Communication, the University of Texas at Austin School of Information, Uppsala University's Department of Informatics and Media, the Department of Aeshetics and Communication at Aarhus University, and the Department of Media and Communication Studies at Södertörn University. I thank Kip Keller once more for helping me find the right words.

Abbreviations

A2K	access to knowledge
ACTA	Anti-Counterfeiting Trade Agreement
ACTN	Advisory Committee on Trade Negotiations
BSA	Business Software Alliance
BT	British Telecom
DMCA	Digital Millennium Copyright Act
DRM	digital rights management
EC	European Commission
EFF	Electronic Frontier Foundation
EU	European Union
EUCD	European Community Directive on Copyright in the Information Society
FFII	Foundation for a Free Information Infrastructure
FIPR	Foundation for Information Policy Research
FOSS	free and open-source software
FRA	Försvarets Radioanstalt
GATS	General Agreement on Trade and Services
GATT	General Agreement on Tariffs and Trade
HADOPI	Haute Autorité pour la Diffusion des Oeuvres et la Protection des Droits sur Internet
IFPA	International Federation of the Phonographic Industry

IIPA	International Intellectual Property Alliance
IMF	International Monetary Fund
IP	intellectual property
IPR	intellectual property rights
IPRED	Intellectual Property Rights Enforcement Directive
ISP	Internet service provider
KEI	Knowledge Ecology International
MEP	member of the European Parliament
MPAA	Motion Picture Association of America
NGO	nongovernmental organization
NSM	new social movement
P2P	peer-to-peer
PhRMA	Pharmaceutical Research and Manufacturers Association
PPI	Pirate Parties International
PP-UK	Pirate Party—UK
QoS	quality of service
RIAA	Recording Industry Association of America
SMS	short message service
SPP	Swedish Pirate Party
TPB	The Pirate Bay
TPP	Trans-Pacific Partnership Agreement
TRIPS	Trade-Related Aspects of Intellectual Property Rights Agreement
USTR	U.S. Trade Representative
WHO	World Health Organization
WIPO	World Intellectual Property Organization
WTO	World Trade Organization

Introduction

Move over, Greens. Make way for the Pirates.
—Mary Papenfuss, "Pirate Party Storms German Ship of State"

The Swedish Pirate Party (SPP) broke through from an incipient social movement of male software programmers and file-sharing geeks to formal political representation quickly, with a "free culture" message that is in conflict with the European Union's legal system. The party crystallized from and alongside a "gas[eous]" civil society group called Piratbyrån—the Piracy Bureau (Marcus Kaarto, in Norton 2006), which led protests over the police takedown of Swedish file-sharing search engine, The Pirate Bay (TPB), in 2006. The coincidence of TPB's takedown with the surge of youthful membership in the Pirate Party demonstrates that participation in music and media sharing contains a political ideology with a mobilizing force. Rickard Falkvinge, the founder of the SPP, describes pirate politics as a generational movement oriented toward communicative freedom: "In the 1960s the buzzwords were peace and love. For this generation, it's openness and free speech. This generation has grown up being able to say anything to anybody. Letting ideas battle it out for themselves. And all of a sudden, corporations want to take that away.

And 'offended' does not do their emotions justice . . . I think we are the next Greens" (qtd. in Cadwalladr 2012).[1] Protecting the legal basis of openness and free speech online is a moral and political project for the Pirates. The German Greens, whose politics resembles the Pirates' in certain ways, seem a likely historical precedent for the movement, as Falkvinge acknowledges. In this book, I explore the idea that the Pirates and Greens are filial movements sharing a similar moral consciousness and an explicit ecological project based on moral and economic notions of a commons, or a public domain. The institutionalization of alternative, cultural-commons models for cultural integration, what I have called an Alternative Jukebox, emerged in the 2000s in such places as TPB and Napster.[2]

Like the Greens, the Pirate Party has internationalized in national assemblies and the European Parliament. The Pirate Party's first Swedish representative to the European Parliament, Christian Engström, joined the Greens/European Free Alliance.[3] As the Pirate Party learns to operate in a minority coalition, it must weigh ideological purity against pragmatism.[4]

The Greens reached political maturity nearly a generation ago, securing a voice in parliamentary decision making and promoting ecological thinking in many Western countries. Falkvinge sees similar success ahead for the Pirates: "Looking at the cycles of history, the time is right for a major new political wave. And the Pirate party is in 56 countries now. We had this smash success where we got into the European parliament in just three and a half years from founding. We became the largest party in that election for people under 30, just sweeping the floor with the most coveted demographic. The establishment didn't know what hit them" (qtd. in Cadwalladr 2012). Though the Pirates won 7 percent of the national vote in Sweden in 2006

(C. Edwards 2009, 3), allowing it to send one of its members to the European Parliament (Clarke 2009, 24), the popular press has puzzled over the Pirate Party's successes, having initially branded it as a one-off stunt, or as an intrinsically illegal, but ultimately irrelevant, political movement. Yet the Pirates generate historically recognizable conflicts in popular communication as expressions of classed, new social movement (NSM) politics. Pirate politics becomes more intelligible as a social movement when it is seen how elements of environmentalism and "cultural environmentalism" are connected (Boyle 2008).

Drawing from Habermas (1984, 1987a, 1987b), I argue that pirate politics is uniquely centered on expanding "communicative rationality," taken here to mean shared, intersubjective, noninstrumental forms of sociality. The Pirate Party evokes principles of communicative rationality (free and open conversation) in its policy proposals for free speech, increased privacy, access, and participation, for all varieties of personal online communication. This normative, and counterfactual—because online anonymity is practically impossible to attain—position makes the Pirate Party suitable for study from the perspective of the theory of communicative action and second-generation critical theory. The Pirate Party's NSM characteristics are likewise amenable to analysis by second-generation critical theory, which has looked to NSMs as phenomena supporting the lifeworld-colonization thesis. The Pirate Party is an expression of resistance to lifeworld colonization by Internet regulation.

Privacy is chief among the concerns of pirate politics. File sharing is broadly perceived as private communication online, and as being very little like publishing. The privacy—*integritet* in Swedish—of one's personally identifying information, a value that lies at the heart of pirate politics, is put at risk by the Swedish pro-copyright organization, the Anti-Piracy Bureau and its allies. "After the Swedish government adopted contentious legislation that criminalized file sharing and allowed the monitoring of emails" (Hartley 2009), the Bureau successfully pressured Internet service providers (ISPs) to conduct antipiracy surveillance on their customers. As a result anonymity and privacy online are counterfactual ideals that appeal strongly to technophilic youths who are digital natives. Pirate politics therefore may be considered a variety of cyberutopianism (Burkart 2010, 121–122). This cyberutopianism also covers the legal domains of free speech and access to information and culture.

File sharing is considered a variety of free speech or, more broadly, freedom of expression. For example, emerging prohibitions against running BitTorrent clients, along with "throttling" practices that restrict the throughput of BitTorrent, threaten to violate a perceived right to communicate. The Pirate Bay's search engine for torrents functions like other search engines, by indexing and making links accessible, but nonetheless came under concerted legal attack for its infringing and noninfringing, dual use. Free speech extends to the contents or headers of the packets flowing in and out of a user's ports, which should be handled in a nondiscriminatory fashion, as with common-carrier regulations for telephony. Internet Protocol address blocking, in its existing and proposed forms, is broadly perceived as censorship. And the proposed three-strikes, or "graduated response," policies to ban accused infringers from the Internet are another perceived threat to digital rights, since access to knowledge and information is a pirate value closely related to privacy and free speech.

The first-ever criminal prosecution of web publishers in Sweden, that of the "Pirate Bay Four," was a spectacular show trial that angered young Internet users, unwittingly promoted the

SPP's national campaign for Riksdag (Swedish parliament) seats, and turned its young founders into national folk heroes. Swedes take pride in their long tradition of press freedoms and free speech rights, and many besides the Pirates thought that TPB should be afforded the same speech rights as web publishers. As the Pirate Party formalized a political identity for the movement, it parted ways with the Piracy Bureau, a think-tank-cum quasi nongovernmental organization (NGO), and campaigned for new online privacy protections, reduced copyright terms, and the cancellation of governmental surveillance programs and the EU Data Retention Directive. In their popular communication, the Pirates cast themselves as a freedom party or a free culture party. Like the Greens, the party arose from civil society as an NSM with defensive characteristics. Their utopianisms share a grammar and even some vocabulary. The counterfactual normative ideal expressed as the Internet commons, the public domain, or the Alternative Jukebox has equivalencies in both environmental and ecological politics.

In elucidating pirate ideology, I focus on the communications-technology and policy discourses about free culture that permeate the literature supporting the cause of cultural environmentalism. In addition, I examine the sociology of the post-1980s European protest movements that have become intertwined with European "technoculture" (Robins and Webster 1999). I emphasize the complementary moments of identity-based and resource-based analyses and approaches that have developed from European and North American research. Despite its strong presence in pirate politics, I do not linger on identity politics, except as historical background to cyberliberties activism.[5] Instead, I focus on the mobilization of resources (information, art, messages, and money) for litigation and public education, and the "mobilization of bias" (Schattschneider 1988) for lobbying and transforming private complaints about cultural suppression into public issues.

My study responds to the fear of the rapid digital "enclosure" of the Internet commons, which would lead to a regulatory dissolution of the public domain. Growing calls from academics for an environmental movement for culture—for example, that by Boyle (2008)—indicate that such a dissolution is already underway in the United States. Boyle presents as a watershed moment in cyberliberties the U.S. Supreme Court's decision in *Eldred v. Ashcroft* (2003), which found constitutional the retroactive extension of copyright terms. Boyle deemed it "the widest legal restriction of speech in the history of the Republic . . . proceed[ing] without significant First Amendment review" (247). Other voices in U.S. academic and legal discourse, including Yochai Benkler, Lawrence Lessig, Siva Vaidhyanathan, William Patry, Kembrew McLeod, Jessica Litman, Christopher M. Kelty, Joanna Demers, Patricia Aufderheide, and Lewis Hyde, express similar concerns about the need for copyright reform to protect and defend the creative commons online. Of these, Boyle's call for a social movement dedicated to cultural environmentalism breaks most clearly with the orthodoxies of legal studies on cyberliberties and may even anticipate interventions by critical legal theory. Similarly, McLeod's Situationist critiques of "copyright bozos" (McLeod 2001, 2005) have intervened in the juridical discourse of intellectual property rights (IPR) by framing ludicrous pranks carried out by McLeod and others to demonstrate concretely some of the absurdities and legal fictions of unreformed copyright law. Boyle does not go further into an analysis of cultural environmentalism and NSMs, and so provides an opening for such a project.

The intellectual-commons discourse in law and policy literature shares an ethic with public-sphere social theory and communication research on net activism. The ethic is democratic
socialist in its treatment of the market, and libertarian-anarchist
in its treatment of individual rights. There is a U.S. bent to this
libertarianism. Both the commons and the public sphere enjoy,
or should enjoy, certain exemptions from systemic pressures. On
the one hand, neither should be subject to unmodulated market
forces. On the other hand, neither should be subject to state-
sanctioned censorship or prior restraint on speech (with gen-
erous latitude given for determining what is protected speech).
Another common feature of public-sphere and intellectual-commons discourses of cyberliberties is the imputed telos of democratic activism. In *Common as Air*, Lewis Hyde spells out why
activist approaches to intellectual property (IP) should focus on
politics and social action. Law and politics are turning "to the
old idea of 'the commons' as a way to approach the collective
side of ownership" (Hyde 2010, 12) and advocates and politicians are intervening earlier in decision making related to "how
we treat art and ideas once they have entered the public sphere"
(18). Hyde's commons shares critical features with Boyle's "public domain," and—as I argue—these form anchors to the media
lifeworld. The public domain, which is largely unpenetrated by
market or political systems, underpins social action promoting
more collective forms of media ownership and access. In Sweden,
a strong conception of the public domain, cultural and natural
commons, and social democracy all hang together (Andersson
2011a).

The public domain of greatest significance initially to pirate
politics was TPB. The SPP and, later, the German Pirates coalesced
around a coherent platform and formal (juridical) identity only

after they weathered a watershed collective event in Europe, the temporary closure of The Pirate Bay's file-sharing site (Hartley 2009). TPB began operations in 2003, two years after Napster was shut down by court order. Despite TPB's relative youth, the cultural significance it has developed for European Internet users can hardly be overstated. TPB continues to operate on servers in Sweden, running on free and open-source software that lets users search for items online without hosting a centralized software tracker. TPB does not host any content on its servers, but instead tracks active BitTorrent sharing services on the Internet. The site's owners were prosecuted and convicted in what many Swedes considered a show trial; they had begun paying penalties at the time of this writing. The site survived, demonstrating itself to be "the world's most resilient bittorent site" both technically and legally (TPB 2003), and so achieved mythological status in the technoculture.[6]

Fan reactions to TPB's temporary closures have a precedent in cyberliberties activism. Popular-music fans and artists, culture jammers, hacktivists, and indie labels in the United States joined in a shared sense of loss and outrage after the "Napster watershed" (Burkart 2010) and expressed conflictual collective action.[7] When The Pirate Bay site was raided by police in 2006, the furor among its millions of users contained enough energy to launch a single-issue party based on the asserted freedom to share files online. Since 2006, the single-issue movement has become institutionalized, grown into a sizable party in Germany as well as Sweden, and articulated a broader agenda that resonates among young people around the world who were born digital: "Pirates believe in direct democracy, transparency in government, online data protection, the legalization of soft drugs, and the implementation of a basic minimum income for all. 'We want more freedom,' said a candidate simply. The Pirates have won the support of many because they're fresh, authentic and funny" (Papenfuss 2011). The hybrid socialist-anarchist platform, which is always evolving from one pirate scene to the next, avoids some of the U.S.-style libertarian characteristics of North American cyberliberties activists. Yet with their North American counterparts, European youth culture shares a lifeworld that is partially constituted in online communicative action. And it directs the shared sense of moral outrage, alluded to by Falkvinge, at technocratic politicians and other decision makers who are out of their depth with the Internet and digital media, but who nonetheless impose harsh restrictions on the use and enjoyment of digital culture.

As Europe rushes to adopt and implement policies even harsher than U.S.-style regulations and cracks down on file sharing, communal outrage over the rapid loss of part of the digital cultural commons is being transformed into collective action and even institutionalized forms of protest. The debates over cyberliberties in Europe have evolved beyond the standard frame cultivated by lobbyists, who claim that although file sharers may use grandiose rhetoric to justify their illegal activities, they just "want stuff for free" (Andersson 2011a, 11) and "steal" private intellectual property to get it. Pirates respond: "It's not theft. It's an infringement on a monopoly. If it was theft and it was property, we wouldn't need a copyright law, ordinary property laws would suffice" (Falkvinge, qtd. in Cadwalladr 2012). In Europe, the incommensurability of pirate politics and established law is debated, since consumer rights to access, privacy, and free speech online, along with antitrust statutes, have an inherited basis in law and compete juridically for standing alongside IPR owners' rights. It is therefore possible to observe pirate politics working within legal, political, and cultural structures laden with possibilities for social transformation.

Alternative modernities are social visions that are built into the outlook of cyberculture. My last book, *Music and Cyberliberties*, provided an empirical study of an incipient movement in the United States and Canada that is elaborating a normative framework for considering the politics of file sharing independently of the received juridical framework for IPR in U.S. or Canadian law. I presented the case that file sharing is such a widespread and accepted practice that it can be said to contribute significantly to cultural reproduction,[8] in addition to providing vital marketing functions for the culture industries, so the political and economic terms of cyberliberties were still negotiable. I characterized file sharing as a technology practice that requires cyberliberties in order to survive—principally, the rights of citizens to privacy, access, and free speech. Pirate politics updates these arguments for a European context.

I also described familiar modes of collective action that are gathering strength in defense of cyberliberties, including hacktivism, culture jamming, alternative and radical media, and media pranks. Most of my examples were taken from North America, and I did not devote much attention to European activism in its various forms, including *netzpolitik* (in German) or *nätpolitik* (in Swedish). In the interim period, the pirate parties of Sweden and Germany staged dramatic entrances onto the European political scene, and a popular discourse arose about the pragmatics of, and emerging opportunities for rolling back U.S.-style regulations under a banner of pirate politics (see, for example, Pickard, in Andersson 2009a). Whereas cyberlibertarian activism is still occurring in "submerged networks" (Melucci 1989) in the United States, where it is not yet institutionalized in the political system, the parliamentary systems in Europe afford the Pirates unprecedented visibility as a maturing social movement.

Cyberliberties Activism

This book extends the argument of *Music and Cyberliberties* by addressing the emergence of the SPP and other national pirate parties as a symptom of damage caused by technocratic controls on private life online and hence on cybercultures. It addresses a positive dialectic too: the expressions of communicative rationality in the emergence of pirate politics. It is informed by a structural-functionalist perspective on social change that addresses the strains and load-bearing capacities of social systems, and also recognizes the social agency and collective identities of movements as change agents. The SPP expresses the cultural shock of "lifeworld structures" (Schutz and Luckmann 1973) confronting expanded markets, bureaucratic-technical processes, and police powers. The Swedish and Scandinavian cybercultures found favorable political and cultural terrain for collective action oriented to preserving the legal regime that had protected TPB from regulation for a few years. Those cybercultures' expectations about sharing cultural objects online were threatened by the "Europeanization" (Fossum and Schlesinger 2007) and integration of transnational communicative spaces juridically.

"Europeanization" is a term that can mean separate cultural and social integration processes at work, at varying speeds, among European Union (EU) member states. There are different uses of the term "Europeanization" in political science and legal discourses (Henderson 2009). I use it and "Europeanizing," provisionally, to mean regionally centralizing and bureaucratizing legal and financial processes that shed nationally organized systems in favor of EU and European Commission (EC) systems. (The EC is the executive body of the EU.) Pirate politics is generally opposed to Europeanizing information policy, as mass protests over the Anti-Counterfeiting Trade Agreement (ACTA) and the Intellectual Property Rights Enforcement Directive (IPRED) in 2011 and 2012 illustrated. It is these and other varieties of conflictual social action that *Pirate Politics* takes as contemporary case studies.

This politics of symbolic action is integrated at local, national, and transnational levels. Beyond Europe, pirate politics carries currents of antiglobalization and transnational contention in its activities in the technoculture. The Pirate Party builds directly on the backlash against the globalization of free trade and IP agreements. That counterblast began to coalesce in opposition to the proposed Multilateral Agreement on Investment (1995–1998) and took visible form at protests at the World Economic Forums, beginning in Seattle in 1999 and recurring since then. Those precursor conflicts inspired new media researchers to develop ideas about collective action in cyberspace and suggested that net activism could be oriented to antiglobalization ends.[9] Likewise, the mass demonstrations worldwide on February 15, 2003, against the George W. Bush administration's aggressions in Iraq set precedents as collective expressions of "complex internationalism and transnational contention" in their mobilization online and in the streets (Tarrow and Della Porta 2005). Something similar, but on a smaller scale, occurred with TPB's closure and news of the seizure of its servers in 2006. Though coming as a blow to free-culture proponents in the wired world, the seizure mobilized bias in favor of a formal political challenge. Pirates registered at public rallies via short message service (SMS) micropayments. The initial response began in Sweden, but as the SPP communicated the scale and scope of the Europeanizing project, an internationalization of pirate politics followed. Pirate parties spread quickly to more than two dozen other countries, including Germany, Spain, Austria, Poland, France, Belgium, Italy, Denmark, Finland, and Canada (Clarke 2009, 24; C. Edwards 2009, 3; K. Jones 2007). Informal pirate parties have sprung up in New Zealand, Argentina, and in Utah (K. Jones 2007) and Florida in the United States.

Pirate politics visibly influences European deliberations and debates about cyberliberties. In 2010, the Pirate Party's representative in the European Parliament led the opposition to the EU Executive's plan to implement ACTA by administrative fiat (EurActiv 2010). ACTA debates in Sweden, the UK, Germany, France, and other countries were mirrored in waves of online and street protests against secret negotiations by EU ministers, renewing concerns about a lack of democracy in EC processes. In February 2012, thousands of anti-ACTA protesters marched in Germany, Poland, the Netherlands, and Sweden. Germany delayed signing ACTA after the justice ministry voiced concerns in the wake of the protests. Germany in fact followed Latvia, Poland, the Czech Republic, and Slovakia in delaying ratification (BBC News 2012). The EP finally rejected ACTA in July 2012. But before ACTA's eventual implementation in the United States, Canada, and Japan, the protests led to open debate and deliberation for the first time on the risks of ACTA to cyberliberties at local, state, and transnational levels.

It is difficult to exaggerate ACTA's significance for mobilizing support for pirate politics. Cyberliberties groups around the world criticized ACTA as a lopsided IPR treaty dressed up as a free trade agreement, and protested ACTA as a global censorship initiative foisted on uninformed Internet users around the world. Health advocates claim it is a stumbling block for generic-drug manufacturers, which would otherwise diffuse off-patent pharmaceuticals more quickly and broadly. New legal liabilities making ISPs responsible for infringing activities pushed smaller ISPs, along with the Pirates, to lobby against ACTA. ACTA's U.S.-style criminalization template for EU member states raised the question of its suitability in Europe and prompted discussions about the relative superiority of different national privacy regimes to the EU proposals. The standardizing of police surveillance powers was especially contentious. But public information about the proposed treaty was scarce throughout its negotiation period.

> The treaty calls for the creation of an "ACTA committee" to make treaty amendments, for which public or judicial review are not required. . . . The panel would also operate outside of the scope of the World Trade Organizations [sic] or the United Nations. However, industry groups would have "consultatory input" to amendments. A 2009 Freedom of Information request showed that Google, eBay, Intel, Dell, News Corp., Sony Pictures, Time Warner, and Verizon had received copies of the draft treaty under a nondisclosure agreement. (GCN 2012)

ACTA's secrecy fed conspiracy theories and Euroskeptics' accusations of overreaching by Brussels. Leaked drafts of ACTA published by WikiLeaks revealed substantial and growing pressures on national ministers to "harmonize up" to the highest protection standards for property owners while eliminating rights of appeal and exemptions for citizens.

Many parliamentarians felt helpless to resist ACTA. Pangs of buyer's remorse hit EU officials following its signing. In January 2012, shortly after the EU acceded to ACTA, Kader Arif, the rapporteur for ACTA in the European Parliament, quit his post in protest over the substance and process of the plurilateral agreement. His resignation statement reveals that corporate colonization as a mode of executive decision making leaves members of the European Parliament (MEPs) unaccountable in the global

IPR-reform program. The pathos of Arif's denouncement communicates that even among ACTA's early supporters in the EP, the means and ends of the program are seen to have become illegitimate.

I want to denounce in the strongest possible manner the entire process that led to the signature of this agreement: no inclusion of civil society organizations, a lack of transparency from the start of the negotiations, repeated postponing of the signature of the text without an explanation being ever given, exclusion of the EU Parliament's demands that were expressed on several occasions in our assembly.

As rapporteur of this text, I have faced never-before-seen manoeuvres from the right wing of this Parliament to impose a rushed calendar before public opinion could be alerted, thus depriving the Parliament of its right to expression and of the tools at its disposal to convey citizens' legitimate demands.

Everyone knows the ACTA agreement is problematic, whether it is its impact on civil liberties, the way it makes Internet access providers liable, its consequences on generic drugs manufacturing, or how little protection it gives to our geographical indications.[10]

This agreement might have major consequences on citizens' lives, and still, everything is being done to prevent the European Parliament from having its say in this matter. That is why today, as I release this report for which I was in charge, I want to send a strong signal and alert the public opinion about this unacceptable situation. I will not take part in this masquerade. (Arif, qtd. in LQN 2012)

Though this public appeal came late in ACTA's progress through the EP, it contributed to the rising profile of symbolic protest over cyberliberties. The ACTA masquerade shows that the EP was prepared to build the infrastructure for the European Celestial Jukebox on the U.S. model (Burkart 2010) through administrative fiat.

Pirate politics grapples with such disputes at the code level of European legal and economic integration, in ways similar to the NSMs that politicized genetically modified food, nuclear power, and the introduction of U.S. tactical nuclear missiles into Europe. Pirates are "challenging codes" (Melucci 1995) through which information policy is managed, using traditional and nontraditional forms of collective action globally and engaging new social forces in the process. The Pirates contest the most authoritarian expressions of network power.

Despite its formal participation in state institutions, the SPP acts extraprocedurally to attain certain goals in cyberspace. For example, it took direct action against the state by becoming the ISP for TPB and WikiLeaks (TorrentFreak 2010a). Unconventional allies are converging around pirate politics, including swarms of Anonymous hacktivists training their "Low Earth Orbit Ion Cannon" on the servers of dozens of multinational media corporations and governmental enforcement agencies. Filial groups and projects include Antisec and Lulzsec, which probe for and exploit security vulnerabilities of their targets; Telecomix, the Western technical facilitators of Arab Spring reporting; WikiLeaks; the Julia Group; and the Icelandic Modern Media Initiative, which proposes establishing a model "data haven" in Iceland (Greenberg 2011). The last was inspired by the Swedish Pirates' proposed governance of a micronation, the Principality of Sealand (Libbenga 2007). The universe of the Pirates is porous and ambiguous, but interconnected actions there, in submerged networks and in concert with other activists, challenge authoritarian codes.

A Short History of the Swedish Pirate Party

Although Pirates prefer to avoid using the term, the birth of the "Swedish Pirate Party" was a symbolic protest against the takeover of decision making about the electronic frontier by

Europeanizing regulations. TPB began operating independently in 2004 (Miegel and Olsson 2008, 203–204). Police raided its headquarters on May 31, 2006. The Pirate Party emerged afterward from a loose collective of anarchistic technophiles calling themselves the Bureau of Piracy (Piratbyrån), the inverted namesake for which was the new and official Anti-Piracy Bureau, which set up shop in Sweden to lobby for the enforcement of new infringement laws for sharing online. Widely derided in popular youth culture, the Anti-Piracy Bureau became emblematic of the political mission of newly self-identified Pirates. The Piracy Bureau began as a support group for Internet radio broadcasters (Li 2009) and developed the formal mission and reformist platform of the SPP, which calls for shortening copyright terms, abolishing patents, and rolling back new Internet usage or content regulations.

The Pirate Bureau describes itself this way:

Piratbyrån is not an organization, at least not primarily. First and foremost, Piratbyrån is since its beginning in 2003 an ongoing conversation. We are reflect[ing] over questions regarding copying, information infrastructure and digital culture. Within the group, [we are] using our own different experiences and skills, as in our daily encounters with other people. These conversations often bring about different kinds of activities. (qtd. in Miegel and Olsson 2008, 207)

The Piracy Bureau monitored and reported on antipiracy measures moving through the Swedish Riksdag, the EP, and abroad. Mobilizing activities included "meetings with net activists from other countries. . . , lectures that representatives of *Piratbyrån* have held for students in different universities in Sweden and at conferences both in Sweden and abroad. . . , conferences and meetings they have taken part in. . . , campaigns and events they have initiated and conducted," and street demonstrations (208).

After the takedown of TPB, the emergent Pirate Party used the Internet to campaign for cyberlibertarian reforms and held simultaneous "pirate demonstrations" in Stockholm and Gothenburg (Sjöström 2006). In the party's first year, its membership exceeded that of the Swedish Green Party (Solomon 2006), which was and still is a minority party. Falkvinge claims that "membership in the Swedish Pirate Party tripled in April after the four men behind The Pirate Bay file sharing website were convicted of copyright infringement by a Swedish court" (qtd. in Hartley 2009). The surge of popularity swelled further as unpopular new laws targeting file sharers generated "a wave of revulsion" (C. Edwards 2009, 3).

The Pirate Party is largely a youth party. "According to an exit poll by . . . [public broadcaster] SVT, 12 percent of men and 4 percent of women voted for the Pirate Party [in the Europarliamentary election]. Among those under 30, 19 percent are believed to have voted for the Pirate Party" (Meza 2009, 25). Most members are younger than thirty: "There are a few seniors, by which I mean people over 30, but the bulk is much, much younger. Honestly, if a member of a traditional party looked at our demographic, they wouldn't believe it. We are peaking at ages 18, 19" (Falkvinge, qtd. in Cadwalladr 2012). I argue that this cohort demonstrates a recognizably class-based consciousness of being at risk of "blocked ascendancy" (Gouldner 1985) by older generations of politically entrenched, conservative, long-lived, and digitally illiterate technocrats, who, in addition, presided over the Great Recession of 2008.

By learning to use traditional political tactics, the SPP won early compromises on pending IP regulations threatening TPB. As a result of the demonstrations and debates the party staged soon after forming, "the Justice Minister agreed to negotiate a revision to the law that had made illegal the unauthorized down-loading of copyrighted material"; not surprisingly, "The Pirate Bay Web site was operating again in Sweden within days." The party's success attracted the attention of the establishment: "The major contenders to form the new government, sensing that the public would elect Pirate party parliamentarians in the coming election, changed their policies to thwart the threat" (Solomon 2006). The system shows early signs of learning from and adapting to pirate politics. Compared with the "authoritarian" (Andersson 2011b) tactics of incumbent politicians responding to the demands of the copyright industries, pirate politics appears to be deeply democratic and anti-authoritarian.

A Short History of the Discourse

Pirate politics is similar in nature to "anti-political politics," as Václav Havel once described Poland's Solidarity movement. NSM theory effervesced in the 1980s in the wake of a series of seemingly unrelated events: Solidarity in Poland, western European mobilization against the deployment of U.S. nuclear weapons on the Continent, growing numbers of national eco-logical protests (especially after the Chernobyl disaster), broader participation by women in the economy and society, and the breakdown of some of the USSR's influence behind the Iron Curtain. Second-generation critical theorists who generally follow Habermas's modernization model—including Klaus Eder, Claus Offe, Agnes Heller, Ferenc Fehér, Nancy Fraser, Jean Cohen, and Andrew Arato—developed theories of collective action for the cultural movements engaging in nontraditional politics in civil society across Europe, while noting the rapid structural changes affecting the economies and societies of Europe and the Eastern

Bloc. They offered an updated modernization theory organized around cultural politics, and retained the critical concepts of class and structuration, even while challenging their continuing relevance. I explore similarities and differences among NSM theorists in chapter 3.[11]

Their studies revealed affinities among transnational groups with radical and alternative visions of social life. I blend this approach with the communication-centered, international history of alternative media and media reform movements, a history first analyzed in John Downing's *Internationalizing Media Theory* (1996) and *Radical Media* (2001), and partly updated in the two-volume *Making Our Media* collection (Rodríguez, Kidd, and Stein 2009).

Although cyberliberties topics are becoming more prevalent in cultural studies, popular-music studies, and critical legal studies of IPR, media and telecommunications policies rarely find a way into the discourse, in spite of the pirate parties. Works on policy studies by Milton Mueller (2004), Joanna Demers (2006), Jonathan Zittrain (2008), James Boyle (2008), and Adrian Johns (2010), along with some policy reports, form a cluster of policy research in information policy, IPR law, and Internet regulation.

But these titles maintain a U.S.-centric perspective, as does work by the greatest critics of digital rights management (DRM) and the Digital Millennium Copyright Act (DMCA; a U.S. law), including Lawrence Lessig, Siva Vaidhyanathan, Pamela Samuelson, Patricia Aufderheide and Peter Jaszi, and Tarleton Gillespie. And these authors, while critical, do not offer a normative theory to ground their defense of and advocacy for free software, free culture (or freer culture), and antitechnocracy. They leave it to juridical discourses to find, in critical scholarship, their own self-justification. This is a puzzling situation, since their own work, collectively, shows the bankruptcy of U.S. IPR laws to be practically self-evident. This paradox leads to results such as issuing oft-repeated appeals to reform copyright law, but without an analysis of the political or cultural opportunity structures for doing so. In the United States, that call is readily deflected by Hollywood and telecommunications interest-group politics.

The Canadian scholar Michael Geist documents changes made in the international juridical systems to accommodate Internet regulation and "cyberlaw 2.0" (Geist 2003). Like Lessig and Vaidhyanathan, Geist adopts a radically democratic perspective on the copyright wars, emphasizing participation and access as critical to democratic media institutions. Geist tracks developments in international communications law and policy that touch the EU and Canada, and provides guidance for comparative systems analysis, but he has not yet contributed scholarship that links policy with communication theory. Geist, like U.S. critics of the DMCA, frequently leaves us guessing about the social-agency characteristics of cyberliberties politics that respond to changes in the international juridical systems. I seek to flesh out the question of political agency of international groups with a cyberlibertarian sensibility.

Organization of the Chapters

Chapter 1 addresses cyberliberties activism and the defense of the lifeworld online, connecting those notions to contemporary struggles over media policy, telecommunications policy, and trade policy, discrete areas that are frequently merged in global and European treaty negotiations. The chapter asks why pirate politics should be evaluated by using critical systems theory and NSM theory.

I engage with NSM theory to accommodate the politics of cyberculture. Like the European ecological movement, pirate politics aspires "to be both a pressure group and a moral crusade" and also expresses a fear of "the uncontrolled and irrational use of the resources of this world" (Eder 1996, 159). The ecological movements in Europe exemplified the "social construction of nature" (1) through cultural politics and related it to Frankfurt School discourses about the damage that instrumental reason ultimately does to a human "inner nature." That philosophical discourse of modernity carries over to pirate politics and the perceived need for an environmental movement for culture. Chapter 1 presents the theoretical rationale for the book, which is to place pirate politics within a broader context of the Internet's normalization, colonization, and impending crisis. It looks at pirate politics as a variety of middle-class radicalism, explores its grievances and mobilization processes, and inquires about the significance of Sweden for its emergence.

Chapter 2 explores in greater detail European antipiracy initiatives, their rationales, and the political responses to them. Specifically, it dwells on the EUCD, IPRED, and ACTA. I highlight the expansion of governmental surveillance powers associated with the April 2009 enforcement of IPRED, which removed privacy protections from Internet users and facilitated the conviction and sentencing of the founders of The Pirate Bay website. Those policy changes have contributed substantively to the grievances of the SPP. The chapter also considers the changing policy regimes for DRM, software patents, Internet commerce, trade in IP, and police surveillance, all of which helped shape the institutionalization of the SPP.

Chapter 2 describes the powerful role played by the Office of the U.S. Trade Representative (USTR) and the contribution of asymmetrical relationships between the United States and its trading partners to conditions conducive to the expansion of pirate politics. The content industries directly lobby the USTR and other executive agencies for changes to U.S. trade policy (B. Smith 2010), and European countries are now the targets of significant pressure. Content industry lobbyists wield outsize influence over European parliaments and executive branches, encouraging them to combat file sharing as a cultural norm and to harden policies they consider too soft (Caldwell 2009). That influence demonstrates the pressures facing Sweden and other, smaller EU states to conform to the needs of U.S. software and digital content exporters. Chapter 2 relates the critical role of the Recording Industry Association of America (RIAA), Motion Picture Association of America (MPAA), and Business Software Alliance (BSA) in coalition building in favor of IPRED and ACTA among MEPs and executive staff members.

Chapter 3 develops the idea of communicative rationality for NSMs generally and applies this idea to Pirates' social learning, their hacking the software of state, and their resemblance to other movements. The chapter examines the cultural environmentalism argument in detail, focusing on the essential function of lifeworld defense. It makes the case that pirate politics expresses concerns of a new middle class that is structurally vulnerable and economically blocked from ascendance.

The story of pirate politics is a tale of how alternative cybercultures cannot coexist alongside increasingly hostile and punitive legal, technical, and economic systems. The Pirates' vision of an alternative information age based on the public domain and the intellectual and cultural commons is likely to flourish only in submerged networks.

Finally, the ecological argument of pirate politics is revisited, together with its regressive potential. The history of philosophical romanticism, which is critical of formal rationalism, shows how some varieties of cultural politics that try to recover nature can wind up as reactionary, irrational, and pseudoscientific belief systems, including racism (Eder 1996, 117). Not much has been reported about how the financial support of a key donor enabled the rapid mobilization of political support for the Pirate Party. In Sweden, Carl Lundström supported the TPB in 2003, but had previously donated significantly to the populist "and xenophobic" New Democratic party in the early 1990s (Caldwell 2009). It is worthwhile asking whether the financial participation of an anti-immigration activist disqualifies an incipient party from being considered a new social movement, since NSMs are typically oriented to leftist politics (although not universally; see Downing 2001). But with the SPP's narrowly focused and universal-rights-based claims to cyberliberties, the "Stockholm daily *Dagens Nyheter* is right to call . . . [it] an antidote to traditional kinds of extremism" (Caldwell 2009). It is ironic indeed that initial support for pirate politics and its principled stand for communicative rationality could emerge in part from such a murky bog.

1 Nomads of the Information Society

 Pirate politics expresses a culturally ingrained libertarian senti-
ment, a youthful and geeky attitude toward autonomous uses
of technology, a preference for confrontational symbolic repertoires, and a wonkish enthusiasm for procompetitive policies
and disruptive technologies.[1] Pirate politics is collective action
oriented toward restoring damaged or threatened communication codes and norms within a politicized cyberculture. It contributes to knowledge of the sociology and politics of NSMs
dedicated to preserving the public domain and defending the
communicative lifeworld more generally.

The mainstreaming of some of the basic messages of pirate
politics suggests that its efforts have been successful. The EU
telecommunications chief, Viviane Reding, stated, "In my view,
growing Internet piracy is a vote of no-confidence in existing
business models and legal solutions. It should be a wake-up call
for policy-makers" (qtd. in Willis 2009). MEP Guy Bono applied
the Pirates' critique of ACTA to an economic critique of the
culture industries that sponsored that legislation: "The repres-
sive measures are measures dictated by industries that have
been unable to change their business models to meet the needs
imposed by the information society. Switching off Internet

access is a powerful sanction which could have profound repercussions in a society where access to the Internet is a mandatory law for social inclusion" (qtd. in Schroeder 2009a).

These remarks by policy makers serve as a reminder that imposition of the Celestial Jukebox is negotiated and not an inevitable outcome. The current uncertainties surrounding the codification of the European Celestial Jukebox emerged from the new politics of file sharing rather than from content deregulation, new competitors, or disruptive business models. Having met little or no real political resistance in the United States to their creation of a draconian copyright regime and the closing down of Napster and Grokster in the 2000s, the RIAA, MPAA, BSA, and rights agencies are still struggling to make sense of the pirate politics that they encounter in Europe. Could the pirate parties become a serious impediment to reforming Europe's information policy, or are they merely a temporary fit of pique? Are they preparing to engage the system formally as a policy doppelganger, or to drop out of the system and stage political theater in strictly symbolic, oblique, and marginal political protests? Will the industry have to negotiate with software pirates and privacy activists?

Theoretical Rationale

In keeping with the theory of communicative action, a theory of pirate politics should frame the problem area as the colonization or normalization of the Internet, define social agency as visible resistance to colonization, and offer a basis for disproving the proposition that pirate politics is an NSM for cyberliberties. Empirically, it identifies the symbolic and functional centrality of file sharing among moderns who participate in cybercultures as evidence of social integration, and activism around sharing as an ideologically middle-class movement.[2] And a theory of pirate politics should adopt NSM theory rather than cultural studies or political economy, since NSM theory is compatible with the theory of communicative action, whereas cultural studies and political economy both encounter unique problems of agency and social change within their own perspectives.

Pirate politics joins the European NSMs of the 1980s in promoting communicatively rational social programs in the face of colonization and crisis. It inaugurates a period of decolonization of the Internet to "save" the Internet from irreversible spoilage. The colonization of the lifeworld online became recognizable only after researchers in publicly funded U.S. labs (Castells 2010) began using the Internet as an open platform, and thus diffusing its networking capability. The Internet's early period of flat network architecture, openness, and equity among "peering" nodes and their communication protocols is seen by contemporary jurists as a fleeting moment (Lessig 2002; Zittrain 2008). The demise of the peering relationships between ISPs, the introduction of metered services and packet discrimination using QoS (quality of service), packet-sniffing surveillance, Internet Protocol address blocking by commercial media services, and the stabilization of cyberlaw have, in about a decade, repurposed the Internet from a sharing platform to an almost ideal commercial-media distribution platform—leaky, but normalized and suitably tamed for business. The Internet normalization thesis of Zittrain and others articulates many of the same processes and consequences as the public sphere desiccation thesis of Habermasian critical theorists.

Resistance to normalization is not unexpected. But an analysis of political agency requires more than merely labeling a political phenomenon a movement; agency requires collective action and an observable dynamic in conflict with colonization processes, social structures, and other, more powerful agents. For every claim that a social movement lies behind a social conflict with a visible expression, there should be an effort to disprove the corresponding null hypothesis—namely, that the expression is explainable by other means, or is a mirage, or is an artifact of analysis. The null hypothesis regarding pirate politics correlates with the party line advanced by the content industries: pirate culture is a manifestation of degraded mass-consumer culture, on par with street riots and looting, and any evidence of a political movement informing the politics of file sharing is, at best, a clever public relations stunt or, at worst, solely the expression of the meanest of economic desires by lazy freeloaders who just "want stuff for free." The null hypothesis postulates that activists are self-interested egoists and that their file sharing is uninformed by reflection on social values or their vulnerability to discovery and punishment under existing law and policy.

The first indication that pirate politics is not just about getting stuff for free came when the SPP mobilized quickly in the midst of a rapid transformation of the Swedish copyright and information policy regimes. The party grew exponentially in its first three years: "Early on, the party garnered a lot of publicity, but only managed to assemble 0.63 percent of the overall votes in the September 2006 election for parliament. In the 2009 European Parliament elections, however, they assembled 7.13 percent of the votes, and gained one seat in the European Parliament" (Andersson 2011a, 24).

Another indicator is the formalization of Kopimism. The moral fitness of file sharing is at the heart of pirate politics as a doxa. The ascendance of Kopimism to the status of a recognized *what if it is?*

religion in Sweden illustrates the broader social significance of pirate politics.[3] Kopimism is an anticopyright belief system that promotes file sharing of all kinds. Its central tenet is no belief in a supernatural being, such as God or gods. It is an implicit religion exhibiting "key traits of religious belief-ritual and performing a similar role as religion without the focus on the divine/transcendent" (Campbell, personal communication). Or, Kopimism may be just another way to rub the nose of the establishment in the formalization of pirate politics. Either way, Kopimism undermines the null hypothesis.

File sharing, for the purpose of this book, can be understood as a politics of symbolic action (Edelman 1971), communicatively meaningful social action. It can be a type of performance that communicates in-group or intersubjective solidarity and political arousal. In the context of movement politics, file sharing contributes to the cultural commons and its reservoirs for personality development, socialization, and creative work online. TPB is a space for what Yochai Benkler (2006) calls "nonmarket"-based communication, and what I call the Internet's lifeworld online, but what content industries call the Darknet.

But although file sharing is meaningful social action, disclosing intentionality and communicative agency, not all file sharing and media sharing, in and of themselves, are expressions of a social or cultural movement. They are a variety of lifeworld experience shared by more and more people who are participating in media urbanism (McCullough 2006). File sharing is a functional requirement of cybercultures, and cybercultures develop as part of people's background expectations and through what is taken for granted about the usability of the world. They are not (yet) politicized.[4] In urban centers in the developing world, and increasingly in rural areas as well, media urbanism means that

sharing music and movies for free or at a nominal cost is second
nature, and typically unassociated with political resistance. For
example, Ravi Sundaram (2010) writes of the "pirate modernity"
that is expressed in media urbanism in Delhi amid widespread
economic poverty. As in Delhi, media urbanism in other impov-
erished global cities thrives more or less independently of established law and policy. Cyberliberties activism is activist because
of a conscious rejection of industry propaganda and juridical
attitudes toward IPR, along the lines of what Andersson (2011b)
describes in the Pirates' personal and conscious rejection of the
null hypothesis.[5]

File sharing aside, NSM theory leads to the discovery that
cyberlibertarians, including the Pirates, tend to be middle-class
(or middle-class-aspiring) radicals from the "new class" of elite or
ascendant knowledge experts.[6] Though some may be philosophical anarchists, they are not revolutionaries. Notwithstanding the
Pirates' collective actions (including seemingly radical participation in Anonymous actions), their social action is not similar
to the collective action of the "multitudes," for reasons including the Pirates' visibility and presence, defensive posture, self-
restricting strategies, and conflicted ideologies. Also, as proposed
in chapter 3, the Pirates, unlike the masses, experience "blocked
ascendancy" (Gouldner 1979) by competing elites, through the
substitution of professionalism by regulationism and through
the restructuring effects of the Great Recession of 2008.

I differ from the poststructuralist position (for example,
that offered by Strangelove [2005]) that identifies the activity
of piracy with a foundationally anticapitalist dynamic. In contrast, pirate politics merely tweaks the logic of existing regimes.
The Pirate Party and its supporting movements represent a new
middle-class interest in preserving dominant social institutions,
including capitalism, as well as privileged positions within them.

interesting ✓

This position—that pirate politics is an updated variety of middle-class radicalism—strays from analyses and conclusions offered by identity-based studies of social movements, which see revolutionary anticapitalist resistance emerging through entrenched local resistance to colonization and colonialism (see, for example, Gibler 2008). It differs also from studies of popular communication that infer political socialization and identity formation from fandom, cultural consumption, "prosumption," and so forth, without considering reification. Those approaches, while flourishing, provide no plausible account of social agency, or even of social action, that explicates mobilization for collective action, the identity politics of cyberculture, or colonization.

I argue that although pirate politics is a form of resistance to abusive forms of domination, its emphasis on reforming copyright to promote cyberliberties more broadly is not a call to eliminate private property rights entirely, but to modify them. It pushes a reformist agenda toward property and information law, promotes capitalist competition in its antimonopolistic stance, appeals to bourgeois rights, and is happy to work within the political system to promote its agenda. It is also happy not to work. Two German Pirates in the Berlin organization told me in 2012 that they preferred studenthood, private sector work, slacking, and leisure time to doing political work, and so would prefer to see other parties take over the cyberliberties agenda than to campaign for office and to govern.

With the threat of a null hypothesis looming over much of the work on cybercultures in media studies, NSM theory needs updating to account for collective action on the Internet, especially collective action as symbolic politics about the Internet. Twenty-five years ago the announced "return of the actor" (Touraine 1988) to political theory and labor history stimulated new interest in forms of collective action that do not conform to traditional models of democratic politics. Only some social movements are precursors to minority political parties. The other groups remain in civil society and stage symbolic warfare, typically against nonstate actors, through media campaigns and direct action. Their causes and appeals can eventually be formalized as membership organizations, although affinity groups can function in cyberspace like memberships to generate and signify popular support.

Internet Zapatismo (1994–2006) presented social movement researchers with an early model of Internet activism. Named for the Zapatistas, Internet Zapatismo supported the indigenous rebels in Chiapas, Mexico, by publishing revolutionary treatises online and motivating foreign supporters to lobby the Mexican president to stay the Mexican army from massacring the jungle fighters and their supporters and to release the political prisoners taken in the Lacandón jungle. Internet Zapatismo exemplifies the use of the "boomerang" strategy (Keck and Sikkink 1998; Tarrow 2005), which refers to media messaging for the purpose of generating external and global pressure on local politics. Internet Zapatismo was conducted through the Internet, but was not "about" the Internet or the disposition of the Internet as a zone of freedom; instead, it was about recognizing the dignity of the struggling locals in Chiapas and throughout indigenous Latin America.[7]

Hacktivism updates NSM theory as a symbolic repertoire of conflictual collective action. It is directed against prominent targets in software, finance, security, and media enterprises; since the mid-1990s it has demonstrated political characteristics, including observable examples of collective action. But hacktivism on its own lacks the political agency of a social movement.[8] It is leaderless collective action that oscillates between anarchistic antiauthoritarianism and vigilantism inspired by revenge attacks (see table 1.1). Hacktivist alliances such as Telecomix, Anonymous, Lulzsec, and Antisec, all of which nurture a countercultural bent, try to provoke social change through the popularization of their accomplishments and rivalries with authorities. The "movement" ascribed to the generation of hacktivists before these renegades has been only partly described and theorized by Strangelove (2005), Coleman (2010), and Wirtén (2006), among others. Reports of Anonymous-oriented hacktivist projects in the last five years suffused the news, drowning out news of classic hacktivist clans, including Legions of the Underground, Hacktivismo, Urban Ka0s, Hong Kong Blondes, and Cult of the Dead Cow. Anonymous's famous tangles with Scientology include "Project Chanology," designed in part to rebuff the church's ongoing filtering campaigns for removing anti-Scientology content at Digg and YouTube (Vichot 2012).

Political mobilization, religious institutionalization, class identification, and links to hacktivism are evidence that pirate politics is a movement, however hive-like it may seem within the Anonymous paradigm of hacktivism. But the present exploration of communicative action and testing of the null hypothesis require a search for an extra variable: namely, determining whether the Pirates' symbolic politics of the Internet and collective action exhibits communicative rationality, on its own or in concert with filial groups and campaigns. If pirate politics does so, then it might be categorized with the ecology, feminism, antiwar, antiglobalization, and antiapartheid movements—social movements organized around specific goals that are "in principle of concern to people of any group" (Shaw 2005, 139). It would join movements whose communicative action attempts to decolonize the lifeworld of the overgrowth of coercive administrative power.

Table 1.1
Selected hacks attributed to Anonymous

Americans for Prosperity

Apple

Bank of America

Bay Area Rapid Transit (San Francisco)

Church of Scientology

Central Intelligence Agency

eBay (PayPal)

Gawker Media

HBGary Federal

International Monetary Fund

Lolita City

Malaysia Ministry of Education

Mastercard

Microsoft

Montreal Police

MPAA

Quebec Education Department

Quebec Liberal Party

RIAA

Sony

Stratfor

U.S. Department of Justice

Visa

Warner Brothers

Westboro Baptist Church

YouTube (Google)

Source: Author.

Although the text-centered approach of cultural studies provides interpretive techniques for considering media content, it lacks a sociological regard for communicative action, and evades normative theory. Political economy nurtures a hermeneutic of suspicion appropriate for assessing the decline and deterioration of lifeworld resources, but it cannot on its own assess the communicatively rational potentials of social movements. It also lacks a theory of communicative action with which to detect and critique colonization. The cultural effects of copyright maximalism, for example, are outside its scope, as is a normative critique of those effects. Political economy cannot interpret or articulate the politics of symbolic action underlying the social and cultural responses to the European experience with the Celestial Jukebox. It needs NSM theory for this.

NSM theory offers an ability to assess decolonization and observe social learning. In previous work (Burkart 2010), I used NSM theory to characterize cyberliberties activism as an incipient, not fully fledged social movement, on the basis of its limited mobilization and fractured social agency. Pirate politics in Europe provides an opportunity to update my cautious assessment by seeking evidence of surplus communicative rationality at work in the pirate parties of Sweden and Germany and associated movements. Critical theorists including Jürgen Habermas (1987a, 1987b), Jean Cohen and Andrew Arato (Cohen and Arato 1992), and Klaus Eder (1996), especially, use system-and-lifeworld sociology to evaluate NSMs according to their potential to transform institutional domination into domains of freedom. The emergence of postmaterialist and new middle-class social movements in Western Europe in the 1980s presented NSM theorists from the social sciences with new examples of political

conflicts, claims, performances, grievances, and identities that
often took a stance oriented toward society at large and not
toward the market or the formal mechanisms of state. NSMs were
seen as cultural observatories or laboratories presenting alternate
social visions of modernity, accommodating the political consequences of legitimacy crises, and even healing lifeworld damage
through accelerated or more advanced forms of social learning
(Eder 1996). NSMs became a focus of critical theory also because
of their distance from the political system and their antipolitical
politics, which are performative, rebellious, and ultimately independent of left-right party identification.

There are other advantages of using NSM theory to study
pirate politics. Because second-generation critical theory's NSM
approach is grounded in the theory of communicative action,
it provides a stereoscopic view of the politics of the information society. It elucidates the symbolic and interpretive basis
for its collective action while providing a historical-material-
ist accounting of its conditions. The normative aspects of the
theory of communicative rationality and the critique of colonization can be put to good use in describing the social field in
which pirate politics operates. The pirate movement propagated
a free culture message from the fringe to the mainstream: from
the clubby TPB to the civil society mobilizations of the Piracy
Bureau, to the institutionalization of the SPP, to the gelling of
the Pirate Parties International. The Pirates symbolize opposition to the colonization of online spaces devoted to media and
communication, and their collective action explicitly thematizes
and advocates for lifeworld resources (free speech, privacy, and
access) that provide and enhance communicative rationality. In
its early phase as a social movement, pirate politics mobilized a

variety of collective actors pursuing social change through cultural politics and savvy campaigning. In its party phase, it pursues change through governance and direct action.

The SPP is an NSM dedicated to social learning and cultural decolonization. Its program is to enhance privacy, access to culture and knowledge, and (to a lesser extent) freedom of speech online. Reformed copyright, abolition of the patent system, and enhanced rights to privacy are the three platform planks promoted at the time of this writing (PPS 2012).[9] The party engages the convergent domains of information, communication, and copyright policy in ways that are recognizable from the symbolic action of earlier social movements. As I argue more fully in chapter 3, pirate politics is replete with environmentalisms. An unnamed vice-chair of the German Pirate Party expressed this concern as a fear of destruction of his "living space" and compared copyright maximalism to "the chopping down of forests destroying the habitat of koala bears. According to him, this destruction had got worse and worse in the past decade and it was his aim to stop the government's 'regulation spleen'" (Löblich and Wendelin 2012, 8).[10]

NSMs reveal the emancipatory power buried under ideology and the ruins of the labor movement. Today its spark is modulated and articulated as identity politics, consumer rights, self-help, and other causes that resist unabated capitalist accumulation and growth, but are stunted by false consciousness and other ideological constraints on solidarity. The theory of communicative action identifies an antisystemic pulse in NSMs, but attributes their emergence to new forms of conflict opened by legitimation crises, the ongoing cycles of crisis attendant on the expanding and integrating of capitalist economies, the economizing and rationalizing of firms and other bureaucracies, and the juridification of everyday life. Conflicts are read symptomatically, within the grand narrative of lifeworld colonization.[11]

Political Grievances

In pirate politics, file sharing underpins basic grievances concerning the disposition of privacy, free speech, and access online. Within a decade of the normalization of broadband consumption, and after the verdicts against Napster (2003) and Grokster (2005), file sharing changed from being a merely suspicious activity to one actively policed, regulated, and criminalized worldwide (Burkart 2010; Burkart and McCourt 2006). The status of file sharing changed as the reform of information policy was pushed by the content industries, U.S. agencies tasked with trade policy, and institutions of the EC and EU.

Those changes followed a timeline of the Internet's normalization and the suppression of its radical potentials. When Hollywood lobbied Congress for deregulation of the market for telecommunications infrastructure and services in the mid-1990s, it enticed lawmakers with the metaphor of the Celestial Jukebox. After acquiring the legal and technical infrastructure for the Jukebox, the content industry won many new rights to prohibit bypass of the system and to practice rent seeking. In the 2000s, record and movie companies sought the criminal prosecution of hundreds of thousands of people—initially, U.S. residents whom private investigators reported to the legal departments of media companies for online sharing (Burkart 2010). Since then, the industry has expanded its control abroad, principally through lobbying the USTR for tough bilateral and plurilateral negotiations for IPR reforms that come wrapped in trade agreements. The aims are to fulfill the original vision of "ubiquitous DRM" (Doctorow 2005)—but the growth of the regime has been uneven at best.

Nevertheless, Internet normalization proceeds apace. By 2004, the entertainment industry had begun prosecuting people suspected of file sharing in the UK, Australia, Denmark, France, Germany, Italy, the Netherlands, Finland, Ireland, Iceland, Japan, and Sweden (Andersson 2011a, 161). As chapter 2 proposes, the imposition on European systems of law of legal norms derived from the U.S.-modeled Celestial Jukebox became a de facto condition of information policy with the passage of legislation such as the EU Telecommunications Directive, the EU Copyright Directive, IPRED, and the EU Data Retention Directive. Many Europeans, however, have not yet accepted it in their daily lives. Other reforms, such as the once-defeated ACTA, wait in the wings for future passage.

If the null hypothesis were applied to the characterization of the home legal systems and traditions of Pirates, it might postulate that permissive IPR and enforcement regimes enable and encourage them. But Sweden's legal system eventually fit the mold of the Celestial Jukebox almost perfectly. From the Berne Convention to the Swedish Copyright Act (1960), which imposes fifty-year terms of protection, "Sweden has consistently adapted its legislation to meet the requirements of international agreements" (Li 2009, 292). The fifty-year term is a "major point of contention" for the SPP (292). The Berne Convention and its floor of protections is also reinscribed and interwoven with the ratified Trade-Related Aspects of Intellectual Property Rights (TRIPS) Agreement, which extended copyright to software in

Sweden, and with the World Intellectual Property Organization (WIPO) Copyright Treaty adopted in 1996, which legitimated DRM and created a right of authors to remedy infringements (293–295). Sweden pressed for a single EU patent system when the country headed the union's rotating six-month presidency in 2009 (Meller 2009). Most important for the SPP, perhaps, is Sweden's commitment to EC obligations, especially information policy harmonization (Li 2009, 296). The commitment helped create both the threat and the political opportunity for campaigning for—and winning—seats in the EP.

In the European phase of the normalization of the Internet, the Swedish Pirates led a concerted backlash against the content industry's strong legal program. As a symbol, the revolt was critical for unveiling and illuminating the policy goals of the EC-EU information society program. Since the Swedish government acceded to the strong copyright policy regime crafted by the EC and based on the emergent model of the "informational state" (Braman 2004), the appearance of pirate politics has raised the stakes for lobbyists and their parliamentary supporters.

Notwithstanding the materialization of pirate politics, the EC's maximalist IP agenda seems to be path-dependent and influenced exclusively by lobbyists. To illustrate the capture of the EC by industry, a survey of the negotiating stakeholders attending an EC convention on file sharing in 2010, before the public announcement of ACTA, shows heavy participation by private sector media, telecommunications, and IPR organizations; moderate participation by public sector representatives; and no representatives from NGOs or civil society groups (see table 1.2).

The narrow range of interests represented at negotiations over file sharing means that policy elites block or minimize the opportunities for counterdiscourse about digital rights and

Table 1.2
Representation at the European Commission stakeholders meeting on
file sharing, 2010

Attendee category	No.
Trade groups	
Publishers and rights societies	18
Carriers and ISPs	15
Labels	6
Integrated media	4
Subtotal	43
Governmental organizations	
Directors general	11
National and regional regulators and public sector	4
Subtotal	15
NGOs	0
Total	58

Source: EC-MARKT 2010.

cyberliberties. Yet it is just these hardening institutions of the
Celestial Jukebox that seem to provide agitated Pirates with new
"political opportunity structures" (Tarrow 1994) for building
transnational solidarity and delegitimizing pending EU IP law.
On July 4, 2012, the EP voted to reject the ACTA treaty 478 to
39, leading Pirate MEP Engström to write:

This is the end of a dossier that the Pirates and the Green group in the
European Parliament have been working hard on ever since we got
elected three years ago, and it is a great victory. The victory belongs
to all activists who have been blogging, demonstrating, and contacting
members of the European Parliament to persuade them to vote no. It
was only thanks to the pressure from the outside, from ordinary citizens
deciding to make a stand for freedom, that we were able to win this.
(Engström 2012b)

A confluence of events led to the SPP's early victories. A vote in June 2008 on a controversial surveillance bill in the Riksdag galvanized privacy advocates and file sharers in opposition while drawing the juridical link between free speech (as file sharing) and personal privacy for voters for the first time. The Försvarets Radioanstalt (FRA) law came into force in 2009. "The law . . . allows the National Defence Radio Establishment (Försvarets Radioanstalt—FRA) [a military intelligence service] to intercept all calls, emails and phone text messages crossing Swedish borders" (*Local* 2008). Concurrently, the Swedish IPRED law was being debated before its passage in February 2009. The law allows copyright claimants the right to track down file sharers' personal identities through their IP addresses (see chapter 2). Those events occurred during the course of The Pirate Bay trial, which was mobilizing what would become the SPP's base.

Li (2009) hypothesizes that the SPP's potential influence for reform would be strongest if it used its grievances to lobby for influence as a national party, much as the Green Party of Sweden did in 1994. In that year, the Swedish Greens won representation in all but one regional government, setting the stage for winning seats in the EP in 1995, when they won 17.2 percent of the vote (U. Mueller 1997). As aggressively as the SPP emerged, however, national electoral success did not come as easily or as early to Swedish Pirates as it did for their German counterparts.

Pirate Politics' Defensive Posture

A corollary of the null hypothesis argues that pirate politics in its electoral expression is a fluke. The popular press identified the Pirates' early victories as signaling voters' political exit from mainstream politics through protest votes, particularly in

Sweden (Soares 2009), Germany (EuroNews 2009; Reissmann 2009), and the Czech Republic (Czech News Agency 2010). Ulf Bjereld expressed this political variant of the null hypothesis: "Many Swedes used the Pirate Party as a protest vote against the established Social Democrats and Moderates. It was a way of registering their discontent without having to vote for xenophobic agendas" (qtd. in Soares 2009). The failure of the SPP to build from its early electoral successes nationally supports that basic proposition. Yet it is also possible to argue that voting pirate is a positive symbolic protest message that expresses lifeworld defense as cultural environmentalism.

The Pirates' defensive posture signals resistance to cultural colonization by law and policy, which is a feature of cultural environmentalism. Malin Littorin-Ferm of the SPP's Ung Pirat youth league (for Pirates under the voting age) expresses antigrowth, antisystemic sentiments in explaining how the Swedish government tried to control the Internet by targeting TPB: "We young people have a whole platform on the Internet, where we have all our social contacts—it is there that we live. The state is trying to control the Internet and, by extension, our private lives" (UPI 2009). But votes by young Pirates for a guaranteed protection of their cultural commons also signals an affirmative agenda, and not merely a rejection of an old one. As students, many young Pirates are already sensitized to state regulations, since students' influence over the education system is already institutionally limited.

Cultural environmentalism may characterize an ethic shared by larger categories of voters. A study by the copyright industries provides an illuminating reflection of how corporate leaders decide on a course of action after confronting the realities of pirate politics for the first time. As the German Pirate Party staged national campaigns in preparation for 2011 national elections, media industry analysts wondered how things could have gone so wrong in Sweden, where two Pirate MEPs were elected. "The problem for the content industry was that the general population in Sweden held anti-IP sentiments," wrote the Property Rights Alliance (Ingdahl 2010, 65). That societal characteristic was due to the lag between the establishment of a flourishing cyberculture in Sweden and the legal imposition of file-sharing controls. Swedes had lived for too long without social institutions organized to promote the rapidly expanding property interests of transnational media, telecommunications, and software companies. And younger Swedes had become completely indoctrinated into pirate politics.

The Property Rights Alliance treated the mere appearance of pirate politics as an existential threat. Its report compares pirate politics with socialism (Ingdahl 2010, 66) and proposes yet more mandatory "public education" about the superior ethics of enhanced, private IPR, similar to the aggressive reeducation campaigns launched in the United States (Gillespie 2009). Of course, those lessons do not invite debate or critical discussion of IP laws or their presumed legitimacy. It is easy to understand the public hostility that this distorted proposal engendered. Majid Yar argues that such "attempts to moralize intellectual property rights and criminalize their violation make recourse to a range of repertoires of justification that attempt to naturalize a capitalistic conception of private property"; in such discourses, "proprietary claims over intangibles are promulgated as an individual right and a social good; their violation is presented as socially harmful and unfair" (Yar 2008, 619). In other words, the Property Rights Alliance works to present strident propaganda as socially sanctioned beliefs in lesson plans delivered through public education systems. But reeducation programs in European public schools seem to have little chance of meaningfully changing attitudes about the disposition of IPR in cyberspace, since, as Rickard Falkvinge says, "There's a complete disconnect between the way the younger generation understands technology and the way the older generation does" (qtd. in Cadwalladr 2012). Demographics alone pose a challenge to would-be corporate educators bent on prosecuting a U.S.-style culture war through lesson plans. Yet it is hardly surprising that the industry takes this route. As Yar notes of U.S. efforts, "The focus of anti-piracy campaigns has fallen in particular upon young people as, in keeping with more general social anxieties about youth delinquency, children are seen as especially problematic" (2008, 619).

That public education, like the legal system, is a terrain of struggle over the normative disposition of file sharing suggests strongly that both sides in the struggle over pirate politics recognize the strategic value of the lifeworld for their objectives, and of digital media and education as portals into young people's consciousness. Yar decries the "ease and effectiveness with which the copyright industries have co-opted the [U.S.] educational system as a collaborator in (re)educating children about copyright and the ongoing attempt to mobilize parents as agents of surveillance and disciplinary correction" (2008, 619). But it is not at all clear that the co-optation of European public education will be as easy or effective, in part because of the ascendancy of pirate politics as an attractive ideology among older students.

Pirate politics' "defensive" stance toward the lifeworld does not mean that it is not strategic, or that it always reacts and never anticipates. Defense of the lifeworld online has expanded to cover areas of policy making beyond copyright and varies from party to party. As mentioned previously, the Swedish

Pirates' agenda began as a single-issue cause: rolling back digital copyright terms and penalties for infringement. The single issue inevitably involved others. The reform of copyright laws meant addressing privacy concerns arising from the online detection of copyright infringers; thus, privacy and the right to remain anonymous online became part of the party's platform as well. And in trying to participate in ongoing decision making about IPR reforms, the Pirates discovered the need to promote public access to secretive and closed negotiations, such as the European Commission Stakeholders Meeting on File Sharing. Engström summarizes an updated position of the SPP: "Our manifesto is to reform copyright laws and gradually abolish the patent system. We oppose mass surveillance and censorship on the net, as in the rest of society. We want to make the EU more democratic and transparent. This is our entire platform" (Engström 2009a). To drive the point home, as ACTA was being debated in secret, Engström introduced a crowd-sourced Internet Bill of Rights to the EP (Masnick 2009b). The Pirate Party of Canada included net neutrality among its goals for 2010 (PPCA 2011). Dozens of permutations of the basic template, most reflecting local concerns and opportunities, show up in the platforms of other pirate parties.

Mobilization of the Swedish and German Pirate Parties

The self-referentiality of pirate politics—Internet activism about the Internet's enabling resources—is an underlying feature of all pirate parties. Symbolically, it is a case of the messenger being the message while sending the message of its messaging. It can be difficult to explain this to those not socialized in cybercultures. Pirates successfully socialize conflicts over Internet regulation that would otherwise be privately internalized, restricted to individual contexts, or lost in translation.[12] The Pirates use sustained campaigns of claims making, a wide array of claim-making performances, and concerted, collective showings of supporters' "worthiness, unity, numbers, and commitment" (Tilly 2006, 182).

As previously discussed, that message is directed primarily to young adults. At a Pirate convention, Engström identified university students as a key constituency for pirate politics worldwide:

It is the collective consensus of the gathered European Leaders that with the scarce resources of a new founded contender party, those resources must be focused on a well-identified front bowling pin. Statistical data states that election participation has been on a continual down slope for the past decade and a half for first-time voters, while at the same time, the core support for our issues are in the 18–30 age range. This data is supported by membership demographics.

Therefore, the identified key catalyst target group is university students. Previous experience from elections where Pirate Parties have participated show[s] that we are unusually strong at technical universities; up to ten times the national average. We need to broaden this scope to all universities. Universities are ideal in that they are a concentrated recruiting ground with people who are generally passionate about what they take part in. (Engström 2012a)

The explicit focus on membership recruitment from technical universities as a mobilizing strategy reinforces the young male geek characteristics of the SPP. The German Pirates claim greater success in promoting women to leadership positions, although women are underrepresented in overall membership.[13]

Pirates rely almost exclusively on the Internet's quasi-anonymous and interactive messaging forums (for example, Internet Relay Chat, blogs, Twitter, and wikis) to communicate with the "shoal of fish" making up their fluid membership structure (Löblich and Wendelin 2012, 12) and to generate an ongoing media spectacle related to free culture. For example, the Slovakian Pirates conduct Facebook-based membership campaigns (CTK 2009); the Pirate Party of Canada cybersquats on domain names of hostile campaigns and publishes pro-pirate material there (Schroeder 2010); TPB announces that it will sell itself for $7.8 million (Schroeder 2009b)—a deal that never transpires; the Pirate Parties International organization adds the Czech and Slovak Pirate Parties (Deutsche Presse-Agentur 2009) and the Finnish Pirate Party (Agence France Presse 2009) to its membership rolls; TPB posts obnoxious blog messages to the RIAA (Schroeder 2010); and thirteen pirate parties and the Electronic Frontier Foundation (EFF) launch an initiative to sue and recover damages for the noninfringing users of MegaUpload after the FBI's shutdown of the site in 2011 (Pirates de Catalunya, n.d.). But besides the incessant campaigning and news-grabbing activities, the Pirates also stage theatrical street demonstrations, which they capture in images and video and blog commentary for use to further mobilize support. For example, images and commentary on the SPP's simultaneous "pirate demonstrations" in Gothenburg and Stockholm in 2006 attracted mainstream media attention while also finding wide redistribution in the techie blogosphere. The Swedish Pirates' multimodal protest style inspired other pirate parties to mix physical and online performativity and to heighten interactivity and participation. Citing John Pløger (2010), Craig Jeffrey writes of a "reciprocal relation between web-based communication and specific 'local' performances of civil action" (2012, 5). For the SPP and other pirate parties, and their filial groups, "youth mobilization often involves both canny use of the media and an ability to orchestrate specific 'presence-events': visceral encounters 'on the street,' for example, that can then be broadcast to a wide variety of young people across the world" (5). Combining face-to-face manifestations with online propaganda creates "variegated spatial networks [that] stand in marked contrast to the relatively stable and 'local' debating society or coffee house of civil society in its 19th-century mode" (5). These examples update the concept of the "mind bomb" Downing (2001) describes in *Radical Media*. The pirate mashup of virtual and three-dimensional sociality gives rise to convivial hackerspaces as well.

For all the interest Pirates generate in their numerous and kaleidoscopic media campaigns, providing a stable web presence and reliable Pirate updates is also valuable public relations. For its more formal political communications, the SPP uses highly interactive websites, whereas its competitors' tend to be merely static, informational sites that restrict user-generated content. Moreover, compared to competitors' sites, the Swedish Pirates' sites boast rates of user participation and interaction that are higher and longer lasting (A. Larsson 2011).

Similarly, the German Pirate Party, the Greens, and the Free Democratic Party conduct highly interactive political campaigns, using social media including chat rooms, online boards, blogs, and wikis, "to foster an immediate exchange with their supporters and to coordinate their offline campaign" (Schweitzer 2011, 319). Offline and mixed-context campaigning includes "flashmobs, petitions, class action suits and conferences" (Löblich and Wendelin 2012, 13). As part of their commitment to interactivity, the Pirates' wikis and blogs facilitate citizen journalism, featuring reports on local and national developments regarding lobbying efforts, impending legislation, police actions, and other news of interest to cyberlibertarians.

Web-based forums break with "the classical top-down orientation" of traditional political campaigning and make "more efforts than their competitors to open their websites for genuine interaction" (Schweitzer 2011, 318). Interactivity is consistent with the pirate parties' horizontal organization and reflects "the high Internet affinity among their respective members, their commitment to Internet politics, the fairly decentralized party structures and their strong internal party democracy" (318). But by sometimes airing dirty laundry in public, web forums can undermine the appearance of a party's unity of purpose.

The German Pirate Party's emergence and mobilization responded to perceived threats to cyberliberties emanating from Brussels, especially in the areas of software patents, data retention, and online censorship (Löblich and Wendelin 2012). German *netzpolitik* matured in public debates about the passage of the EU Data Retention Directive (Directive 2006/24/EC) in 2006, which was also the year of the German Pirate Party's founding. In 2009, conflict erupted over the Access Impediment Act, a Domain Name System-blocking law that was seen as using child pornography as a pretext allowing the foes of TPB to block that site and others. The German Pirate Party developed a successful meme against the act (*Zensursula*, a portmanteau combining the German word for *censorship* with *Ursula*, the first name for the sponsor of the act, Ursula von der Leyen) that popularized the cause with ironic wordplay. The Bundestag passed the act in 2009, but, owing in large part to public opposition, it failed to be implemented and was withdrawn in 2011 (EDRI 2011).

In 2009, the first year the German Pirates put forward candidates for office, they conducted a "Freedom not Fear" demonstration in Berlin, which drew about 25,000 participants (Löblich and Wendelin 2012, 13). They won seats in two local elections, but none in the Bundestag. But in several hacktivist pranks, German Pirate Party supporters "leaked" farcical exit polls on Twitter showing the Pirates with "impossible" double-digit vote percentages, to the consternation of election officials (Carter 2009).

The same year, the German Pirates benefited from an endorsement by a founding member of the German Green Party, Herbert Rusche (Rusche 2009). In Berlin, the German Pirates staged mass demonstrations with broad antisurveillance themes in preparation for national elections. The fifth annual (2011) "Freedom not Fear" rally in the Alexanderplatz gathered 5,000 people (including many "people in black") in a techno-music-suffused march against surveillance and expansive new data-retention laws (Wyrembek 2011). Participants included Pirates, Trotskyists, Young Liberals, Antifascists, Unionists, 9/11 Truthers, and members of the Green Party—"the usual suspects," according to observers of Berlin's regular political rallies (Schweitzer, personal communication). Some paraded with the Guy Fawkes masks adopted by Anonymous, while others hoisted cardboard video surveillance cameras above the crowds. Protest signs reading "Build the police state" and "Put cameras everywhere" sent ironic messages. In 2011, just five years after the party's founding, the Pirates won 8.5 percent of the votes cast and fifteen seats in the Bundestag. Elections during the first half of 2012 saw twenty-six new parliamentary seats taken by Pirates in two states.

The German and Swedish Pirate Parties crossed thresholds for formal political recognition, but typically for marginal parties in parliamentary systems, they did not take power. As oppositional parties of a purist bent, the Pirates flirt with power without seizing it, an ongoing reminder that loyal opposition to the Celestial Jukebox remains vigilant, just outside the gates. Their continual activism and campaigning in civil society, which highlight the boundaries of system and lifeworld, depend in large measure on playing an outsider role while policing the margin. The Theory of Communicative Action, better than agonistic accounts of social movements (e.g., Mouffe 1999) discloses a politics of recognition by the Pirates consistent with new social movements, especially in their ascription of universal interests to robust digital rights. As an identity politics of middle-class radicals, the linkage between particular and universal interests exists in cyberliberties, and demands recognition and coexistence. Agonism is expressed as opposition to exclusion from a social means of realizing identity online, as chapter 3 discusses in greater detail.

The Pirate Parties International

The development of the Pirate Parties International (PPI) as a clearinghouse for pirate party campaigns worldwide indicates that pirate politics carries global appeal and that the scope of the conflict over cyberliberties has become international and public. The proliferation of pirate parties supports a claim by Donatella della Porta and Sidney Tarrow that transnational activism adapts to technocratic trends in Western democratic countries.

Internally, there has been a continuing shift in power from parliaments to the executive, and, within the executive, to the bureaucracy and quasi-independent agencies. Power has moved from mass-parties to parties that have been variously defined as "catchall," "professional-electoral," or "cartel" parties . . . and therefore from party activists to the "new party professionals." *Externally*, there has been a shift in the locus of institutional power from the national to both the supranational and the regional levels, with the increasing power of international institutions, especially economic ones (World Bank, International Monetary Fund [IMF], World Trade Organization [WTO]), and some regional ones. (Della Porta and Tarrow 2005, 2; emphasis in original)

These trends constitute Europeanizing practices of trade harmonization. Transnational collective actors, including the PPI, respond to these trends by code shifting from local to regional and global levels as the situation demands. Code shifting allows confederates to communicate lifeworld boundary violations consistent with repertoires of contention established in previous pirate campaigns. Della Porta and Tarrow describe "coordinated international campaigns on the part of networks of activists against international actors, other states, or international institutions" (2005, 2–3), such as those seen in global justice movements and peace movements. The dozens of parties in the PPI work similarly, but also seek votes in local, state, national, and EP electoral campaigns.

In its collation of global bases of cyberliberties activism, the PPI illustrates what Mitchell Cohen (1992) first called the "rooted cosmopolitanism" of transnational social movements. The PPI was founded in April 2010 at a conference of pirate parties in Brussels (PPI n.d.). Its organizational structure is ad hoc and amorphous, and membership is free. In 2012, the PPI comprised over sixty national representatives from political parties from around the world, although few of them had formally registered at home.

The organization gelled into an international collective action against the FBI on behalf of those who had been using MegaUpload's file-storage service for noninfringing purposes. The FBI shut down MegaUpload in a coordinated police action with New Zealand in 2012, and the PPI set up an online submission system to take an inventory of aggrieved parties and their claims. The PPI announced: "[The] Pirates of Catalonia, in collaboration with Pirate Parties International and other pirate parties, have begun investigating these potential breaches of law and will facilitate submission of complaints against the U.S. authorities in as many countries as possible, to ensure a positive and just result. This initiative is a starting point for legitimate internet users to help defend themselves from the legal abuses promoted by those wishing to aggressively lock away cultural materials for their own financial gain" (TorrentFreak 2012a). With assistance from the EFF, the PPI is researching the liabilities incurred by the authorities in possible violation of Spanish electronic privacy laws, among others. In a separate action, the PPI helped mobilize street protests against ACTA in more than two hundred cities in 2012 (TorrentFreak 2012b), with thousands of demonstrators donning the familiar Guy Fawkes masks.[14]

The PPI's wiki and web pages are resources for fledgling pirate parties. Its open publication system permits online content creation and editing. A multilanguage forum for networking, identifying resources, and sharing news and rumors, the website represents the transnational presence of like-minded cyberlibertarians who are working the spaces, connections, and opportunities created by the initial successes of the SPP.

The PPI, German Pirate Party, and SPP challenge social codes that make up the dominant vision of the European information society. They mobilize support by mixing physical demonstrations with permanent online campaigning, viral media stunts, and other low-budget, unorthodox approaches. They contribute to a European web of "networked publics" (Varnelis 2008) in the margins between the political economy and interpenetrating cybercultures.

As mentioned previously, the theory of communicative action and its complementary NSM sociology evaluate social conflicts as expressions of stresses and strains of modernization and colonization. A disadvantage of an unadorned Habermasian approach to NSMs is that it may treat transnational mobilization efforts as symptoms of colonization and alienation without addressing their potential for communicative rationality. An argument by Klaus Eder and Maria Kousis offers to amend that standard reading of the cultural and social significance of the radical politics of collective action. Specifically, they consider environmentalism a mechanism for fostering the evolution of European society through policy discourse and popular debate (Eder and Kousis 2001, 26). Elsewhere, Eder discusses "cultural movements," including environmentalism, as having the "potential to realize the vision of another modernity" (1996, 137). If pirate politics resembles ecology in being an ascendant social movement with an alternative vision for a more communicatively rational society, then it is worth revisiting the purely symptomatic reading and exploring the cultural environmentalism proposition more closely; this occurs in chapter 3.

Why Sweden?

Political economy and policy analysis reveal that the Pirates are not a reflexive expression of economic crisis or deprivation. The Swedish Pirates, like the state from which they hail, advocate imposing political limits on commodification and spatialization. They self-limit, reflect, and debate rather than pushing total or absolutist claims.

Sweden was ready for pirate politics for several reasons. Sweden's international political economy of communication reveals key information-society characteristics. A mesh of services and applications embeds the Swedish technoculture in a communicative lifeworld online. Put another way, the telecom and software infrastructure provides something akin to a dial tone for Swedes born digital (although cell phone users have, for the most part, stopped hearing telephone dial tones). Moreover, Swedish and other Scandinavian software developers have contributed heavily to the code base and user communities of "convivial technologies" (Illich 1973).[15] Its legal system supports pirate politics, and its cultural norms with respect to the natural environment contribute potentially meaningful features to the cultural commons. Those features enabled a political sensibility about everyday uses and enjoyment of the Internet and computers to emerge early and vigorously. Sweden's communication infrastructure underpins technology practices such as private file sharing using darknets, secure e-mail, media blogs, Pastebin, PiratePad, Usenet, open peer-to-peer (P2P) systems like BitTorrent, anonymous storage lockers, and torrent directories. And Sweden's hacker subcultures, hackerspaces, hospitable environments for media and software entrepreneurs, gift economy for locally produced music, and distinctive legal theories support these technology practices.[16]

High levels of connectivity provide protective support for Sweden's knowledge economy. Wireless and wireline broadband are nearly ubiquitous, and a well-developed technoculture is oriented to the digital distribution of music, media, software, and documents. The sharing culture, infrastructure, and literacy in convivial technologies contribute to a robust media lifeworld. Knowledge work and symbolic work contributes strongly to the economy, particularly in cities. The state subsidizes not only public media but also private publishers (Weibull 2003) and increasingly promotes competition in broadcasting markets (Rønning 2003).

The country performs well in high-tech design and manufacturing also. In 2005, 44.4 percent of GDP came from exports of goods and services, principally from "engineering products" (OECD 2008a; see table 1.3). That percentage had increased to about 50 percent of GDP by 2011 (GlobalComms 2012). The country's most important trading partner is Germany, "which took 10.2 percent of exports and accounted for 17.9 percent of imports in 2009" (GlobalComms 2012). Sweden was Germany's fourteenth-largest recipient of exports in 2011 (Statistisches Bundesamt 2012).

Sweden's historically mixed economy and social-democratic parliamentary government (together known as the "Swedish Model") allow it to avoid some of the political and financial crises that would have obliged the country to open all system inputs to free trade and, consequently, to free-flow information policies. A crisis in 1991 precipitated political compromises with traditionally strong trade unions and a break with protectionism and some welfare-state supports. It set the stage for the national election in 2006 of a center-right government led by the Moderate party. As it is, Sweden's economy selectively participates in and stands apart from international and regional integration.

In foreign trade, it overexports the products of its knowledge-based industries and is a net uploader of new creative products to the world. Yet it resists fuller Europeanization by rebuffing currency integration.[17] Habermas calls such a stance against the

Table 1.3
Exports from Sweden, 2006

Category	%
Engineering products	48.8
Chemical products	12.2
Wood and paper products	11.4
Other products	10.9
Mineral products	10.5
Energy products	6.1
Total	99.9

Source: OECD 2008a.

single market a feature of Euroskepticism (Habermas 2001, 95), although Sweden is often compared to Germany and Norway as a European safe haven in the long economic downturn. By not adopting the euro, Sweden has avoided some of the "transmitted" risks from banking and fiscal crises in Greece, Spain, Italy, Portugal, Ireland, and Cyprus (although it risks overvaluing its currency).

When discussing pirate politics at Södertörn University with Swedish faculty and graduate researchers of digital media, law, and policy in 2011, a participant from Sweden declared that Swedes "love the state" for its social contract, rather than distrust it or reject it as an influence in private economic life. In Sweden, participation in entrepreneurship, software start-ups, and competitive media markets for export is subsidized by public programs such as free or low-cost health insurance, retirement and education benefits, and other benefits, which, though threatened, have been retained in Swedish public life. The state offers a minimum monthly payment that can support creative workers, including musicians, open-source programmers, and others, with a safety net. German Pirate Party officials have pressed to have the basic-income guarantee instituted in that country, to better remunerate and provide better incentives for musicians, free-software developers, and other creators (Urbach, personal communication). This net has frayed somewhat since the 1990s.

While the pursuit of export-led growth has weakened Sweden's ability to protect home industries, the country regulates to maintain national champions in strategic industries such as banking, manufacturing, media, and telecommunications. Sweden uses economic policy and financial independence to build in buffers around its engines for growth. The government still owns substantial portions of Nordea bank, TeliaSonera telecom, OMX exchange, and SAS airline carrier (WordIQ 2010). Swedes impose political limits to commodification in their domestic market through regulation, tax policy, and environmental policies, among other ways.

The socially planned Swedish economy features high employment, social welfare, and economic independence (Vihriälä 1991). While Sweden does not have a large economy by regional standards, and is not as rich as its neighbor Norway, the country has grown at a stable rate since the banking bust and currency crisis of the early 1990s, which followed a debt crisis in the late 1980s. After those crises, the country implemented a tighter fiscal policy and strong banking regulations, in the process taking baby steps to modify the traditional Swedish economic model. It persists in its basic form today, despite ongoing reductions in taxes and welfare benefits since the 1990s (EIU 2011). The Pirates' cyberculture rests on a comfortable economic cushion.

It rests also on a carefully planned infrastructure. Sweden has had a distinguished "ICT [information and communication technology] for development" policy centered on broadband connectivity since the late 1990s. Swedish lawmakers made priorities of ubiquitous broadband connectivity and universal service. Achieving them was seen as a "crucial step on the way to the information society" when "broadband penetration was still below 5 percent" (Eskelinen, Frank, and Hirvonen 2008, 415). Sweden publicly funded its broadband network, whereas Finland, which has a similar degree of connectivity, promoted competition and private capitalization (420).[18] Between 2000 and 2005, Sweden spent SEK 4,150 million, or approximately €400 million, on broadband-infrastructure development, including running fiber to local facilities, developing a national backbone to compete with the commercial backbone, and subsidizing "operator-neutral" regional and local networks (416). The measures were designed to be procompetitive, to discourage upstream and downstream gatekeepers, and to speed up "the rate of market development" (417) in web content and services by emphasizing infrastructure build-out and universal service. Information policy and copyright are separate layers of commodification and spatialization resting upon this liberal infrastructure design.

Pirate politics as a political attitude about free culture is nurtured as much by political grievances as by accessible infrastructure, somewhat competitive markets for telecommunications services, and large, centralized institutions. Pirate politics depends also upon a consumption norm of broadband use, mobile and fixed. Household broadband penetration in Sweden is 74 percent (GlobalComms 2012), and availability exceeds 95 percent (Eskelinen, Frank, and Hirvonen 2008, 415). Most household broadband accounts are bundled together with multichannel digital television (GlobalComms 2012), providing greater opportunities for intellectual property "leakage" from TV to the Internet. Sweden is considered to be especially vulnerable to IP leakage since it has the "highest volume of direct fiber connections of any country in Western Europe" (GlobalComms 2012). Table 1.4 presents a snapshot of Sweden's penetration rates for broadband Internet and telecommunications infrastructures. The figures suggest that Sweden's universal service policy, an example of social regulation exceeding the requirements of the EU Universal Service Directive (Directive 2002/22/EC), is working as expected, based on take-up. Wireless broadband data access is pervasive. Mobile operators report that more than 96 percent of the population can have 3G service, and there were 4.1 million mobile broadband users in 2010 (GlobalComms 2012). Young adults, especially, cherish their mobile phones, such that the relationship becomes "perpetual and personal" (Axelsson 2010). Sweden's overall connectivity has matured to the point that its rate of growth has leveled.

Table 1.4

Wireless telephony and broadband Internet connectivity in Sweden

Wireless subscribers	13,810,000
Population penetration	146.10%
Quarterly growth	1.0%
Broadband subscribers	3,015,000
Household broadband penetration	74.10%
Quarterly growth	0.40%
Total PSTN lines (2011)	3,155,000
Household penetration	74.10%
Annual growth	-11.70%

Source: GlobalComms 2012.

Note: PSTN = public switched telephone network

Swedish communication history features Ericsson, the transnational telecoms hardware vendor that develops technology for mobile and fixed networks, television and media, mobile handsets, and business services (Forbes 2012); and Telia, Sweden's largest network operator. Ericsson and Telia together "dominate the Swedish IT industry" (Johanssen 2004, 272), but a large population of Swedish software engineers participates broadly in the thriving market for developing free and open-source software (FOSS). Internet technology development, specifically, contributes 6.6 percent to Sweden's GDP and employs about 60,000 people (Nylander 2011). The ICT components of Sweden's industries are measured at "over 8 percent" by the OECD, which notes that Sweden's information economy fared better than many rivals during Great Recession (2009, 22).[19] Niklas Zennström's Kazaa and Skype distinguish contemporary Swedish programming. Arguably, the region's IT luminary remains Linus Benedict Torvalds, a Swedish-speaking software engineer from Finland who coded and released the Linux kernel in 1994 and works at the Linux Foundation. Locally developed "free" services such as Spotify (for music streaming)—whose programmers honed their skills by designing uTorrent, one of the world's most popular BitTorrent clients—BayFiles (file hosting), and PasteBay (anonymous blogging) are interwoven with the noncommercial repertoire of alternative services and with commercially developed services (such as search, YouTube, and webmail) that have local and internationalized versions available in Sweden.

As mentioned previously, TPB's emergence, operation, defense, survival, and resilience as an online service sustain pirate politics in the social activity of online sharing. TPB's voluntary and ad hoc organization promotes a supportive environment for FOSS and social-software development around open and hybrid services and provides a hybrid search and social networking service.

More generally, Sweden nurtures alternative political scenes that cultivate Internet networking as a mode of organization (Olsson 2008), and many Swedes do take an oppositional and hacktivist orientation toward the culture industries and their technical handmaidens (Andersson 2011a). It is not surprising that in spite of the liberalization of Sweden's information infrastructure policies and the country's openness to trade in software and media, the Swedish technoculture is suspicious of, and hostile to, the cultural and political terms of the Celestial Jukebox. Evidence of this orientation is provided by the emergence of a counterhegemonic "Swedish Model" for media production and distribution (Baym 2011). For example, many Swedish musicians have come to evade the major labels and, instead, publish and distribute digital music through fan blogs, digital syndicators, online affinity groups in social networks, and live events, creating a cultural commons of which fans and a confederacy of indie acts are in charge. The Swedish model of the music industry permits greater participation by smaller and newer acts than a major-label-dominated market does, but also potentially restricts blockbuster acts and ownership consolidation among record labels. Pirate politics and the Swedish Model for music, taken together, suggest that the Internet's normalization in Sweden is incomplete and inconsistent, despite Sweden's commitments to IP Europeanization.

Like file sharing, the Swedish Model for music arose in part as forms of DRM such as regional encoding on DVDs spread to IP blocking and "region filtering" by Netflix, iTunes, and other online content providers, restricting viewing to countries authorized for streaming performances or file downloads. So, for example, Netflix came late to the market with a small catalog, and many iTunes catalog items are unavailable in Sweden because of licensing problems. Thus, cash-rich customers there are locked out of the same portions of the U.S.-managed Celestial Jukebox as the cash poor who reside at the edge of the world's networks and are blocked from legal access to software and media by licensing and the digital divide. DRM regionalization compounds the access problems brought about by region-filtering characteristics such as poverty, low connectivity, and the digital divide. As many people who live outside the United States know from experience with computers, web portals, smart phones, e-readers, and other devices, file sharing and the circumvention of copy protection are frequently the only ways to access software and digital media that are licensed for U.S. consumption but are locked down with DRM for much of the rest of the world. From an orthodox media-economics perspective, such as that shared by decision makers at the EP, these are market failures that can be resolved with improvements to "single market" reforms and IP harmonization. From a pirate perspective, access controls and technical countermeasures to piracy are a first-order social problem to be resolved by legally eliminating them and adopting alternative business models.[20]

Features of the Swedish legal environment favored the emergence of pirate politics, too. Despite plaintiffs' desire for The Pirate Bay Four to be tried in the United States, international IP law specified that Sweden host the trial: "Under the Berne/TRIPS international copyright protection regime, the law of the country where the infringement allegedly takes place applies. Therefore, Swedish law, not U.S. law, governs the dispute" (Touloumis 2009, 259). As it happens, Sweden's legal system is "open ended," meaning that it can be more participatory than other civil law systems, allowing for "groups representing the public, who are not essentially represented by the judicial language of lawyers," to be "invited to participate" by using "understanding-oriented standards" and procedures (Carlsson 1995, 476). Bo Carlsson finds potentials in Swedish law for using understanding-oriented, procedurally grounded public debates to decide environmental and health cases. For example, Sweden's National Environment Protection Board, unlike more formal legal systems, in a limited way accepts moral-practical arguments and exhibits a "procedural rationality" that evolves to recognize new types of legal claims (Carlsson 1995). The Swedish legal code accommodates enhanced consideration of social and political factors beyond the specific claims in a case. For all those reasons, the legal rationales and defenses regarding pirate politics may have received a fuller hearing in Sweden than they could have received elsewhere, despite the outcome of The Pirate Bay trial. The open-ended legal system permits a broad hearing of grievances against the imposition of new restrictions and reduces the chance of their being imposed by executive or administrative fiat. In short, the system uses more communicatively rational processes and can facilitate social learning.

Finally, the SPP's recent emergence as a political movement after legal attacks on TPB is linked culturally and historically to a medieval history of borders, migration, and community. The anti-private-property norms espoused by political pirates, with respect to IP laws in cyberspace, are congruent with the anti-private-property norms espoused by common Swedes in their traditional practices of freely roaming the Swedish countryside and exploiting open and public access to private land, lakes, and waterfronts. The pirates look at the Internet's cultural resources as communal. Rather than thinking of online culture as being a question of "my" file, "their" server, or "our" network, the boundaries between mine, ours, and theirs are blurred.[21] A modern example of anti-private-property rights— the Swedish open-access tradition—links to contemporary conflicts emerging in cyberpolitics and indicates that terrestrial open-access laws are also strained by economic integration with the EU.

Culturally, the pirate identification of cyberspace with public rights of access and privileges to private estates is linked to the traditional Swedish practice of *allemansrätten*, or right of public access known as "every man's right" to "roam and camp in the countryside" (Chu 2009). "Anyone, resident and tourist, can walk [or ski] freely on anyone's land, pick wild berries, mushrooms and non-endangered wildflowers, swim and boat in any waterway, and camp anywhere without a landowner's permission for one night" (Shiffer 1992, 20). At the same time, the right requires that a property owner's personal space not be invaded and that property not be despoiled or destroyed. That traditional environmental norm, which is incommensurable with private property law, has a symbolic extension in online file sharing and support for free culture. Karl Palmås argues that *allemansrätten* is "the opposite of 'No Trespassing'" (Chu 2009). James Boyle construes "the opposite of 'No Trespassing'" as a challenge to capitalism that signifies "the opposite of [private] property" in the public domain (2008, xv).

Shiffer claims that *allemansrätten* is defended more strongly than free speech, religion, or even private property in Sweden (1992, 19). *Allemansrätten* owes historically to a medieval recognition of the right to survival. Members of a small and dispersed population had to travel great distances between towns, often in extreme weather. Society afforded all members the liberty to travel, gather food in the forest, and rest for the night (20). As a "traditionally developed usufructory right" it promoted open access, non-exclusion, and non-rivalrous uses of land (Mortazavi 1997, 609).

The ancient tradition led to the establishment of provincial laws in the thirteenth and fourteenth centuries that established the basis for the modern right. Therefore, *allemansrätten* did not develop as a national law, but as a normative principle and "old habit" owing to "the character of the growth of civilization in Sweden" (Colby 1988, 254). It informs Swedish sensibilities about sustainable development at a foundational level (Swedish Environmental Protection Agency n.d.).

This legal tradition is widespread: "Norway and Sweden not only allow free rambling across private land; they also permit fruit-picking, riding and skiing" (*Economist* 2004). The English public's rights of "rambling" onto private estates on established trails have been expanded to broader and broader portions of the UK over the years (Economist 2004). Rambling and *allemansrätten* describe for the physical landscape what free culture describes for cyberspace: decriminalized surfing and noncommercial home copying of digital media stumbled upon by a home user, student, researcher, or artist. Online territories seem even more limitless and oriented toward freedom.

As urbanization and tourism intensified in Sweden, land-use pressures in the countryside caused public agencies to promote a corollary notion, *allemänsskyldighet*, or the public duty and obligation of everyone to take care of the land (Shiffer 1992, 20). Swedes resisted codifying *allemansrätten* for fear of weakening its normative value as a national moral code for land ethics rules (Colby 1988; Mortazavi 1997; Shiffer 1992). Since Sweden joined the EU in 1991, foreigners, especially poorer Eastern Europeans, have had the ability to harvest berries and mushrooms with the same equal access to open land use as nationals, leading some Swedes to express concerns about unfair advantage being taken of the open-access tradition (Mortazavi 1997, 612).

The null hypothesis is undermined substantially by the foregoing considerations. Pirate politics questions the legitimacy of the Europeanizing information policy regime at the level of policy analysis and formal political communication, and at the identity level, by making information politics personal to (mostly) young people. Habermas's colonization thesis provides an explanatory framework for how this is possible. The political economy of communication further explicates the process institutionally, as factors conditioned by local, regional, and global markets for information goods and services; preexisting and new alternatives to these markets; Europeanizing communication infrastructure law and policy; and existing alternatives to these markets. The Swedish technoculture and legal culture inform pirate sensibilities and the Swedish Model for media. Pirate politics is a variety of cyberliberties activism that has adapted its message to appeal globally from unique circumstances. It remains to be seen whether it will fulfill the expectation by Boyle (2008) that "cultural environmentalism" can become a global human rights movement based on copyright reform and an information policy oriented toward tolerating noncommercial sharing.

2 European Antipiracy Initiatives: The Ratchet and the Fulcrum

A sign above the exit of the autonomous zone of Christiania (Christianshavn) in Copenhagen, Denmark, reads: "You are now entering the EU." Inside Christiania, which originated in 1971 as a squatters' camp in abandoned army barracks, locals developed an intentional community that was distinctively functional while remaining radically alternative to city life in surrounding Copenhagen. Communalism mixed with anarchistic respect for autonomous living—and a thriving counterculture—made Christiania a regional curiosity and eventually an international tourist destination. As property values in Copenhagen rose, developers pressured the city to police and privatize the community's territory. Christiania became caught up in a long-standing struggle between the lifeworld and the system. "Live Christiania" appears across the city on stickers and clothes as evidence of popular resistance to law-and-order social conservatives and the incessant pressures put on the community by local moneyed interests.[1]

Christiania is symbolic of the fate of the not-yet-colonized Internet across Europe. Antiauthoritarian and countercultural virtual communities still thrive, as does Christiania, even under growing pressures of colonization. Even if counterculturalism on the model of 1960s and 1970s hippies is now an artifact, the

Euroskepticism expressed in Christiania's exit signage resonates with the digital natives of Generation X, Generation Y, and the Millennial Generation, whose online Christianias are targeted for extinction. In Sweden, especially, the home of TPB and WikiLeaks, the gap between the implementation of various European directives and the norms of file sharing reveals a crisis of legitimacy in the legal system governing artistic and creative works.

Law and policy can help explain an otherwise inexplicable state of affairs: just as masses of people are said to reach the technologically and culturally enabled phase of the "prosumer," the restrictions placed on their ability to share, remix, and play with copyrighted digital cultural objects in popular communication have increased to the point that the prosumer is likely to be a copyright criminal. The theory of communicative action explains how the system replaces and disempowers action oriented to mutual understanding, by substituting market mechanisms and bureaucratic routines for deliberative decision making. This chapter explores the growth and "harmonization" of the system (and its political, legal, and economic subsystems) through bilateral, multilateral, and plurilateral trade agreements for IPR. These features of Europeanization are most relevant to pirate politics.

The overall harmonization process is ponderous from the perspective of rights holders that lobby the EU for reforms; yet for digital-rights activists, it is occurring with breathtaking speed. Comparatively, the adoption of Internet regulationism in the United States has been expeditious. The Napster and Grokster precedents in the U.S. Supreme Court, together with other important federal court decisions such as *In re Aimster* (2003), had cemented the reliability of the U.S. digital copyright regime for IP rights holders by 2006. With no major-party support for reforming the DMCA in favor of user rights, the one-way ratcheting up of relief for IP owners' rights and creation and growth of user sanctions has proceeded apace. By comparison, the "transposition" of EU requirements into EU member states' laws has been disputational: fraught with problems of implementation and conflicting interpretations, and contingent on national-level politics (Gasser and Girsberger 2004). However, the transposition is on track to yield an IPR enforcement regime stricter than that of the United States. Harmonization, continually hampered by the fragmented system of national regulations, now faces pirate politics as well. Online Christianias, for now, can still find their niches.

Since information policy is reflexive and path-dependent—that is, "decisions made at one point determine the range of possible further options available" (Braman 2006, 325)—pirate politics operates at a critical juncture historically. New to the field of legislation and treaty negotiations, it could one day help tip the balance of information policy in favor of greater communicative rationality.

Trade policy governing IPR provides an ideal vehicle for colonization of the lifeworld online. Negotiations can involve a great deal of deliberation at first, when new treaties and other agreements are being crafted; however, after their creation and passage, deliberation practically ceases, and the system uses money and bureaucratic power—nonlinguistic steering media—to run on autopilot. Over time, an upward IP "ratchet" (Sell 2010) is formed by the replication and expansion of free trade treaties serving as wrappers for the IPR reforms inside them. The expansion of the market for addressable media consumers is paired with new and stronger IPR and enforcement mechanisms with which to collect royalty payments. The logical outcome of the global IP ratchet is the "pay-per" society, in which every packet of commodified information and digital media is accounted for and any consumption of it requires an exchange of payment or labor (Mosco 1989). Global DRM accompanies the pay-per society, such that the ratchet "rebuilds" the Internet as a universal and perpetual surveillance organ as useful for police as it is for attorneys and marketers (S. Larsson 2011, 8).

The WTO and EC are responsible for negotiating and coordinating the treaties designed to expand the flows of commercial trade in digital media and cultural services. Two tiers of strong players exert the greatest pressure on weaker trading partners to adopt the treaties: the United States, the EC, Canada, and Japan; and France, Sweden, and the United Kingdom, the only three large net exporters of "royalties and license fees" within the EU (OECD 2008b, 25). Sweden, being both a smaller trader and a net exporter of IP, sits in a conflicted zone of influence.

This chapter, like the others, is inspired by a desire to show that the Pirate Party did not emerge from "out of nowhere" (Cadwalladr 2012), popular media reports notwithstanding. System-side political economy demonstrates the structural basis for the rebellious new cultural politics: the trade-related aspects of copyright, trademarks, and patents provide the strategic bases and "reason of state" for U.S. and EU cooperation on Internet normalization. The colonization impetus is expressed most clearly by the existence of the trade fulcrum—the structural power relationships and bargaining power expressed in and through trade.

Common Markets versus the Cultural Commons

Changes to European copyright law are driven in large part by the EU's changing relationships to the United States in the course of the former's involvement in multilateral bargaining (Meunier and Nicolaïdis 2006), and by a continuing drive to forge an internal market.[2] Historically, these reforms—especially IPRED (2004) and the Data Retention Directive (2006), discussed later in this chapter—"update" the Internet to better serve as an interactive platform for television, radio, publications, and shopping, with heavily restricted user rights.[3] They were implemented during a period when digital audiovisual and software content was becoming differentiated across the entertainment industry as European states developed a single framework for handling IP within a growing common market. The underlying rationale for the laws and policies covered here is that the system in place for Internet Protocol television and pay-per-view TV, DVDs, DRM-ed digital music and software, computer games, and other media with built-in copy controls should extend to the practical use and operation of the Internet.[4]

U.S. trade policies on IPR, which are practically dictated by the royalties industry, rely on applying political and economic pressure through bilateral, multilateral, and plurilateral negotiations. Europe's trade relationship with the United States in media, culture, and services related to IP is complex and more competitive than its analogous relationships with the rest of the world. International trade plays "a more important role for the EU than for the US in their respective assertions of power" (Meunier and Nicolaïdis 2006, 192), which may help explain why the Europeans have matched and even surpassed U.S. enthusiasm for operating the IP ratchet.

The New IP Authoritarianism in Sweden and the EU

The IP ratchet carries political consequences. "IP authoritarianism" is Andersson's term for the characteristically harsh mode of policy making pushed multilaterally by the United States,

Japan, Canada, and the EU that promotes the criminalization of private, noncommercial file sharing, ongoing copyright-term extensions, ever-stronger legal remedies for aggrieved IP owners, and the ongoing diminution of consumer rights and freedoms in the digital domain. European trade relationships within the EU and between the EU and the rest of the world ensure the structuration of European society and its legal systems as capitalist in the "E.U. way" (Meunier and Nicolaïdis 2006). Facing continuing trade deficits of six billion to eight billion dollars annually in media and communications with the United States (EC 2005a, 5), the twenty-seven principal countries of the EU responded with IPR laws to promote friction-free trade and investment within the eurozone. Being both a net exporter of IP and a smaller state in the EU economically, Sweden is predisposed to meet the stated juridical needs of digital "content" exporters and, to a very large extent, has done so.[5]

The centralizing initiatives of the European common market and the EU are intended to address the problem of fragmentation, or the existence of multiple national legal systems regulating information and media. In addition, trade liberalization in all "electronic communications" and markets for telecommunications services was designed to promote investment in the communication sector. For the purpose of this study, the Framework Directive (2002/21/EC) set the pace. It "lays down clear and stable rules that should create certainty for investors" seeking access to the common market (EC 2005b).[6]

The political motivation for targeting TPB, along with the origins of the legal reforms used for justifying TPB's treatment, is widely ascribed to the United States, especially by prominent Pirates such as Falkvinge. The role of the U.S. government—enacted principally by the U.S. Department of State and the

USTR—is, perhaps understandably, perceived to be that of a belligerent ringmaster forcing foreign governments to conform to frequently uninvited IPR reforms (Drahos 2002). While the United States exercises outsize influence in the bilateral and multilateral trade negotiations in which IPR reforms are typically negotiated, it is joined by the EU, Canada, and Japan in moving the global IP ratchet upward by harmonizing old reforms and initiating new ones. The EU aligned itself with the United States during the Uruguay Round of WTO trade negotiations, which extended free trade to IP and services.

Since the WTO-TRIPS agreement, for example, IPR reforms have been promoted by the same alliance of Northern Hemisphere governments and their flagship companies in the media, entertainment, pharmaceuticals, and microprocessor sectors. The coalition puts pressure on national negotiators to add new protection and enforcement measures to existing international agreements. The developing world, Pirates, and cyberliberties activists see the trend as one that shrinks the public domain and reduces access to knowledge.

The flourishing of cyberlibertarianism in the face of the ratchet heightens the stakes of pirate politics greatly for both sides of the culture clash, pitting an alliance of U.S.-EU technocrats against a broadening resistance. At issue is the legitimacy of the prevailing, global IPR regime. As recently as 2011, law scholars saw "no signs of deregulation or decreasing [copyright] protection . . . indeed the opposite is the case" (S. Larsson 2011, 27). The appearance of pirate politics notwithstanding, copyright maximalists are "undaunted by recent setbacks at the multilateral level" and "have launched a major, almost surreptitious, anti-A2K [anti-access-to-knowledge] campaign focused on 'counterfeiting,' 'piracy' and 'enforcement'" (Sell 2010, 2). These campaigns fuel pirate politics by stoking a growing concern that new IPR are negotiated secretly and without public accountability. While secrecy in trade talks is by no means new, and frequently benefits the negotiating parties, Pirates claim that it undermines the legitimacy of any new laws and regulations emerging from IPR trade negotiations, because trade in IPR is qualitatively different, and more important, than other kinds of trade. ACTA's defeat in the European Parliament in July 2012 set the clearest boundaries yet for the maximalist agenda in Europe, in many ways complementing the defeats of the Stop Online Piracy Act and the Protect IP Act earlier in the same year in the United States, which saw similar activism on the Internet mobilize the resistance.

The international trade relationship between the EU and the United States ultimately drives the ratchet. The entertainment industry's contribution to the positive U.S. trade balance was $9.5 billion in 2005, which was "larger than the positive trade balance of telecommunications and computer information systems combined and 12 percent of the entire United States private sector trade surplus" (OECD 2008b, 2). This strategically vital sector of the U.S. economy hires lobbyists to represent its interests abroad, partly because the royalties earned in media and research-and-development exports to Europe are kept in a closed loop of intrafirm trade among importers and exporters within the same transnational parent companies. That recirculation of royalties reflects "the desire of companies possessing intangible assets to retain a certain degree of control over those assets" (OECD 2008b, 26). The WTO's General Agreement on Trade and Services (GATS) and TRIPS require nondiscriminatory treatment of media flows from the United States to the EU, within certain limits imposed by member states through cultural exemptions, so that the transnational circuits for recirculating capital and content are treated with "most favored nation" status. Offshore banking strategies can further shelter capital from tax liabilities (Shaxson 2012).

Global markets for media and entertainment are substantial; revenues from the "other services" trade category grew almost 300 percent from 1995 to 2005. The International Monetary Fund classifies global trade in media and entertainment in a variety of ways, including global service exports, royalties, license fees, patents, business services, and cultural services.[7] Table 2.1 takes a closer look at the royalty-bearing "other services" category. This portion of "other services" accounts for the greatest share—$32 billion—of the U.S. trade surplus in 2005. About half of that came from trade with the EU. The EU trade deficit in "other services" in the same year was $16 billion (OECD 2008b, 25).

The trade relationship between the United States and the EU in media and entertainment is mutually dependent. The EU takes substantial flows of U.S. media without becoming entirely dependent on it for programming, while the United States takes a smaller share of media content and services from European sources: "The difference between the two entities lies on the

Table 2.1

Global trade in media and entertainment—other services ($billion)

Category	1995	2005
Royalties and license fees	55	130
Computers and information services	10	75
Communication services	25	50
Cultural services	10	25
Total	100	280

Source: OECD 2008b, 24

Note: Figures are rounded to the nearest $5 billion.

import side, such that the United States shows a surplus (of 9 billion dollars) and the European Union a deficit (of 2 billion dollars)" (OECD 2008b, 26). U.S. exports accrue almost half the world's royalties and license fees for patents, trademarks, and copyrights. Together, the United States and the EU generate nearly 75 percent of the world's exports in cultural services, 32 percent coming from the United States and 42 percent from EU countries. Within the EU there is a regional trade imbalance as well: "Only the United Kingdom, France and Sweden achieve significant surpluses for [other services]" (OECD 2008b, 25).

As mentioned previously, U.S. firms recapture many of the profits generated by recirculating cultural services imported from the United States by European subsidiaries. Intrafirm trade by transnational media and pharmaceutical firms based in the U.S. is responsible for the greatest flow of exports to European countries. Sixty-six percent of all licensed or royalty-earning services exported from the United States to European markets occur as "transactions between US parent companies, which own the intellectual property rights, and their affiliates abroad, mainly in Europe" (OECD 2008b, 26). The royalties siphon reduces the stimulative effect of foreign capital on local spending and investment and has the potential of displacing local creative work from media and marketing slots and channels.

Critics of multilateral and bilateral free trade agreements find the growth in international trade in culture, and the hegemonic influence of U.S. firms, to be evidence of "latter-day impositions of imperial power" (Calabrese and Briziarelli 2011, 392). The proliferation of unbalanced trade relationships around IPR creates a "fulcrum" (392) of U.S. advantage in markets for digital media, although, arguably, those advantages accrue to the

United States and the EU alike, proportionately. The U.S. interests served in the trade relationship with the EU are strictly private, since information policy making is, practically speaking, already privatized and decentralized (Elkin-Koren 2000; *Copycense* 2009).

The IPR trade fulcrum uses state action in trade negotiations to gain market access, via incentives and threats of sanctions, to maximize the influence of U.S. exporters. For example, the U.S. Advisory Committee on Trade Negotiations (ACTN), composed of Fortune 500 business leaders and led for many years by Pfizer executives, provided direct input into U.S. trade policy at executive levels beginning in the 1970s (Drahos 2002, 69, 115), and was instrumental in advocating a "trade based approach to intellectual property" reforms during the Reagan administration (70). The trade-based approach prioritizes macroeconomic advantage over other social values when developing information policy. The ACTN and, later, the International Intellectual Property Alliance (IIPA) effectively captured the USTR and set its agenda of promoting harmonized rules globally for anticounterfeiting, enhanced software copyrights and patents, strengthened enforcement, and the rest of the anti-A2K program. When the USTR put up no resistance, the IIPA and ACTN came to act as "the research and policy sides of a larger collective enterprise" (Karaganis 2010, 3).[8]

The United States uses every opportunity to advance the maximalist IPR program. Drahos describes how U.S. policy alternated for decades between encouraging multilateral organizations to pursue the maximalist IPR agenda and then pursuing it through bilateral negotiations: "After World War II the U.S. had pursued a policy of international forum shifting in order to secure the results it wanted in various international regimes" (2002, 112). In intergovernmental forum shifting, "strong states like the U.S. shift forums to optimize their power and advantages and minimize opposition" (Sell 2010, 4), along the path of least resistance.[9] At various points in history, the United States focused on the WTO, WIPO, and World Health Organization (WHO) in its efforts to operate the IPR ratchet, whereas developing countries preferred to work within UN institutions and with more restricted agendas. The United States found itself blocked by Brazil and India in utilizing WIPO to expand protections for IPR (Drahos 2002, 114), so it shifted to the WTO. The Uruguay Round of WTO trade negotiations were a major victory for the maximalists.

The strong Swedish IPR regime is consistent with the policies of a net exporter of IPR. Pressure to adopt maximalist laws such as the InfoSoc Directive, IPRED, the Data Retention Directive, the Telecoms Reform Package, and ACTA comes from the largest rights-holding groups in Sweden as well as the U.S. government. U.S. governmental pressure was increased during the active phases of The Pirate Bay trial. The IIPA, the International Federation of the Phonographic Industry (IFPI), the Pharmaceutical Research and Manufacturers Association (PhRMA), and the Antipiratbyrån represent Sweden's prominent rights-holders organizations. (The IFPI is an international trade organization comprising 1,400 record companies in sixty countries; the companies had $15.9 billion in global sales in 2010 [Smirke 2011]). They shape and benefit from European copyright reform and, because of the escalating copyfight, are partly responsible for adding to the anxiety of the Swedish bureaucrats who manage the country's trade relationship with the United States. The trade groups work at micro and macro levels to influence enforcement norms for IPR violations. Besides petitioning local courts to identify purported file sharers and unauthorized media distributors, they advise the USTR about the amount of pressure to put on the government of Sweden in the trade relationship by applying the IPR fulcrum. They were especially vocal during Sweden's listing or delisting in the stages of the 301 review process (discussion to follow), which, since 1995, has retained a symbolic power of sanction even after the United States acceded to the WTO dispute-resolution process. The historical unity of purpose of the rights-holding organizations contrasts with a relative lack of solidarity among smaller countries that may benefit collectively from resistance (Drahos 2002).

Of the media and entertainment trade groups exerting pressure on Sweden and other U.S. trade partners to ratchet up IP harmonization, the Antipiratbyrån is the prominently Swedish organization of IPR holders with considerable clout in the Riksdag and in the executive branch. Its members' annual revenues are presented in table 2.2.

With even more clout than the Antipiratbyrån is the IIPA, a lobbying group that describes itself as "a private sector coalition, formed in 1984, of trade associations representing U.S. copyright-based industries in bilateral and multilateral efforts working to improve international protection and enforcement of copyrighted materials and open up foreign markets closed by piracy and other market access barriers" (IIPA n.d.). The IIPA, together with U.S.-based PhRMA, exerts lobbying pressure on the USTR and U.S. Department of State, which in turn put pressure, via U.S. embassies, on Sweden and other countries perceived to be undermining the market advantages of those trade associations. The membership list and revenues of the IIPA is reproduced in table 2.3.

Table 2.2

Members of the Antipiratbyrån

Corporation	Revenue, 2010 ($million)
PAN Vision AB	130.3
AB Svensk Filmindustri	103.7
Nordisk Film AB	56.9
Warner Bros Sweden AB / Entertainment	39.1
Succéfilm AB	37.4
United International Pictures AB	22.6
Twentieth Century Fox Sweden / Home Entertainment AB	16.8
Universal Pictures Nordic AB	13.3
Sandrew Metronome Distribution Sverige AB	10.8
Paramount Home Entertainment Sweden AB	9.0
Noble Distribution Sweden AB (Noble Entertainment AB)	8.9
Swedish Film AB	6.4
Sonet Film AB	6.0
Sony Pictures Releasing Sweden AB	4.4
Scanbox Entertainment Sweden AB	3.9
NonStop Entertainment AB	3.2
Folkets bio AB	1.9
Capitol Film Distribution AB (Capitol Invest i Varberg AB)	1.7
Buena Vista International Sweden AB / Home Entertainment AB	0.7
Atlantic Film AB	0.3
Total	477.3

Sources: Antipiratbyrån 2012, largestcompanies.se, Factiva, European Central Bank euro foreign exchange reference rates.

Notes: Revenues were not published for Triangelfilm AB, another member of the group. The figure for Sony Pictures Releasing Sweden AB is from 2012. Nonprofit members include the Svenska Filminstitutet and FilmCentrum. Antipiratbyrån's partners include the International Video Federation (IVF), MPAA, and Nordic Copyright Bureau (NCB) (Antipiratbyrån 2012). Sources were reported in euros; U.S. dollars were calculated at the 2010 daily average of $1.33 per euro.

Table 2.3
IIPA members

Organization	Revenue, 2010 ($billion)
Business Software Alliance	63.9
Motion Picture Association of America	31.8
Association of American Publishers	27.9
Entertainment Software Association	25.1
Recording Industry Association of America	6.9
National Music Publishers' Association	6.6
Independent Film and Television Alliance	4.0
Total	166.2

Sources: AAP 2012, ESA 2012, IFTA 2012, IIPA 2007, Marketline 2011, MPAA 2010, NMPA 2001, RIAA 2010.

Notes: The figure for the National Music Publishers' Association is from 2001. Organizational reports were used for all member organizations except for the BSA. The BSA figure includes only general business productivity and home-use-applications sales data from Marketline 2011.

Although the United States turned to multilateralism from bilateralism in IPR trade negotiations when it joined the WTO, the country retains its "Section 301" review powers for more than symbolic ends. Originally used as a means of warning trade partners of impending sanctions, and of encouraging preferential treatment of U.S. exports, the 301 trade reviews continue to discipline countries seen as disadvantaging U.S. goods and services. The role of 301 review changed in U.S. trade policy and trade negotiations after the creation of the WTO and its dispute-resolution process, namely, the review process now feeds into WTO-TRIPS dispute resolution: "The U.S. government has learned that any Section 301 retaliation will be a target of countersuit at the WTO and has decided to route most of the section 301 cases through the WTO" (Iida 2004, 216). The IIPA's well-funded offices work as a private agency assisting the USTR's work on various 301 projects. They designate geographic areas of concern annually on the 301 timetables; their list of countries of concern has expanded to include practically all the countries in the world. "Internet piracy specifically, was labeled a 'significant concern' in Brazil, Canada, China, India, Italy, Russia, Spain and Ukraine" in 2010 (Game-Politics 2010). The IIPA recommended that Sweden make watch lists in 2008 and 2009 and for the first time actively monitor its citizens. Sweden's neighbors fared less well by comparison.

Congress's annual "Special 301 Report" on piracy used to focus on Russia, China, and Southeast Asia. This year's report placed on its list of problem countries Finland, Norway, and Italy. It noted that Spain is a place where pirating music and movies over the Internet is "widely perceived as an acceptable cultural phenomenon, and the situation is exacerbated by a government policy that has essentially decriminalized illicit [peer-to-peer] file sharing." (Caldwell 2009)

On his blog, Falkvinge occasionally notes what pleasant company Sweden keeps on 301 lists and reports.

Building the European Celestial Jukebox

IPR treaties are typically ratified only after complex negotiation periods. The modern history of pan-European copyrights began with the Berne Convention for the Protection of Literary and Artistic Works of 1886, which was driven by trade concerns. The Berne Convention formalized the rights of authors specifically and harmonized copyright protections for foreign authors internationally, including a standard level of protection for foreign authors across Europe. Subsequent revisions to the Berne Convention added new IPR protections, but failed to define "infringement" precisely. Berne came to be perceived as insufficient as the scale of publishers and other rights-owning organizations grew. Publishers began demanding enforcement mechanisms and sanctions, which the Berne agreement lacked (Agarwal 2010, 794).[10]

The General Agreement on Tariffs and Trade (GATT) came into force in 1948 as a supporting structure for postwar Western nations, alongside the World Bank and International Monetary Fund. The EU and its member states abide by its "common commercial policy; negotiation and conclusion of international agreements; and association agreements involving reciprocal rights and obligations" (EP 2000). The Uruguay Round negotiations transformed GATT into the WTO. The WTO is at the core of the contemporary global free trade regime, together with GATS. GATS covers communication services, including audiovisual and telecommunications.

The WTO contains a TRIPS section, which came into force in 1995. Sweden joined the WTO in 1999, approving TRIPS in the process. TRIPS broadened the global law for IPR beyond copyright and created an information-policy structure based on commercial free trade principles (Agarwal 2010, 795). TRIPS provided for criminal sanctions for illegal copying and counterfeiting on a commercial scale (Article 61), and stimulated the further harmonization of IP laws among member states. It requires participating countries to "consider certain TRIPS provisions while constructing or amending their own relevant laws," but it does not require them to "alter their laws to be in conformity with *all* of its provisions," such as criminal sanctions (806; emphasis in original), as IPRED2 is intended to require. Sweden delayed implementation of criminal sanctions.

There is also the UN framework for multilateral treaty negotiation. WIPO is a special UN agency working with the WTO
to encourage developing countries to become compliant with
TRIPS. The European Union Directive on Copyright in the Information Society (EUCD, Directive 2001/29/EC; also known as the
InfoSoc Directive) implemented the WIPO Copyright Treaty and
the WIPO Performances and Phonograms Treaty for the EU in
2002 (Gasser and Girsberger 2004, 6). The EUCD's ostensible
aims are to address organized crime benefiting from counter-
feit goods and online piracy, an ineffective enforcement regime
in the eurozone, a fragmented internal market for digital goods
and services, reduced business confidence, and "diminished
investment in creative ventures" (Agarwal 2010, 798). When
the EUCD linked Europe's fate as a competitive trade zone for
information and knowledge to strengthened IPR, national parliaments in member states were, in effect, handed their legislative agendas by the EC. Sweden implemented the EUCD in 2005
(S. Larsson 2011, 14).

In its anticircumvention ban, the EUCD provides a framework similar to that of the DMCA.[11] The DMCA implemented
the WIPO Copyright and Performances and Phonograms treaties in the United States. Yet the EUCD goes further than the
DMCA: it "has meant a wider scope for copyright and a criminalization of more actions" (S. Larsson 2011, 16). In Sweden,
the EUCD expanded protections for DRM from computer programs to music and other media (16). But Sweden's law lagged
the DMCA by seven years, permitting a flourishing of Swedish
hacker cyberculture, in the breach.

For its anti-A2K orientation and effect, the EUCD, like the
DMCA, is "likely to have a profound, lasting impact on the
development and use of copyrighted content and technology

in the coming century" (Herman and Gandy 2006, 123). For Bill Herman and Oscar Gandy, who recognize the path dependency of copyright law, this impact is mainly due to the unilateral extension of property rights for IP owners, which excludes user rights. In Europe as in the United States, the ban on the circumvention of technological protection measures (such as DRM) of digital content offers users nothing in return (123).

IPRED

IPRED (Directive 2004/48/EC) codified the minimal enforcement mechanisms to be replicated nationally across the EU by 2006. Its aim was "to address the growing problem of piracy and counterfeiting in the EU Single Market and to resolve the issue of existing disparities among the intellectual property laws of EU Member States" (Agarwal 2010, 797). IPRED's civil and administrative remedies "provide the IPR holder with redress and mitigate the harm of IP violations" (798). Most important for pirate politics, IPRED enables courts to issue injunctions requiring ISPs to turn over the identities of purported copyright violators to rights holders having some evidence of wrongdoing (S. Larsson 2011, 17). The directive provides for search and seizure orders, injunctions, and the freezing of bank accounts linked to those accused of infringing on IPR. Critics charge that the directive assumes guilt until innocence is proved, since the mere charge of infringement triggers IPRED provisions (Hinze 2004).

As drafted, IPRED criminalized IPR infringement, but as adopted it replaced the criminal sanctions with "civil and administrative sanctions" (Agarwal 2010, 798). The change came because many member states, including the UK, objected to the inclusion of criminal sanctions in a free trade, single-market directive (798). After the criminal sanctions were removed, IPRED had a "relatively trouble-free legislative journey" (799) through the EP, and was adopted after its first reading.

Sweden's late implementation of IPRED facilitated a robust public discussion about alternatives to what were perceived as being authoritarian prohibitions on file sharing. One alternative path not taken was a broadband tax: "Before 2008, several commentators—most notably Peter Jenner in the UK and Roger Wallis in Sweden—argued for an extended 'blank media levy' or 'broadband tax' on private Internet connections in order to extend the current system of collecting societies to also be monitoring and reimbursing p2p-type circulation of copyrighted material" (Andersson 2011a, 158). The redistribution of blank media levies was widely perceived to be a less invasive and more palatable alternative than the enforcement directive for providing remedies for claims of lost revenues from file sharing (Andersson 2011a, 158–159).

According to an analysis by the Foundation for Information Policy Research (FIPR) in 2005, industry stakeholders were divided on the introduction of IPRED and its sequel, IPRED2. Favoring the legislation were the music industry, with its historically close ties to the U.S. government (Kravets 2009), large consumer-product companies, pharmaceutical companies, and automakers (FIPR 2005). The EC sponsored the passage of IPRED, giving supportive stakeholders an edge in influencing the specifics of the directive. Opposition to IPRED came from ISPs, auto-parts makers, makers of generic drugs, and civil society groups ranging from open-software groups to librarians. The varied opposition groups were able to make significant revisions to IPRED and to scuttle proposals for strengthening software patents, but they did not mobilize sufficient resources to derail the entire directive (EFF 2007).

IPRED remains a widespread target of critics who question its basic fairness.

Some criticize . . . [IPRED] . . . as being overly broad in its scope because, despite its purported aim of addressing specific concerns of commercial counterfeiting and piracy, it applies to "any infringement of intellectual property rights." Others have raised concerns about the failures to define critical terms such as "intellectual property" and to distinguish between corrective measures to be taken in cases of willful commercial scale infringement and unintentional noncommercial infringement. (Agarwal 2010, 799)

The original IPRED proposal "was met with massive criticism" in Sweden and abroad. ISPs objected to the data-retention requirements of the directive, finding them to be in basic conflict with laws and policies protecting user privacy. "A number of operators [state] . . . that they discard the identification information that the IPRED directive allows access to as early as possible," and anonymizing services have been offered as alternatives to regular service (Larsson and Svensson 2010, 17).

Only twenty-five member states had implemented IPRED by the 2009 deadline (Agarwal 2010, 813); more widespread passage was delayed by data-retention protests and the early waves of pirate politics. In Sweden, public awareness of IPRED grew as the implementation deadline approached. Upon implementation in 2009, Internet traffic in the country immediately tumbled 30 percent (Fiveash 2009), although the volume recovered once users began adopting encrypted virtual private network services (such as TPB's IPREDator) in order to maintain their privacy.

As mentioned previously, ISPs joined pirates, auto parts, and generic drugs makers in resisting the IPRED. The ISP TeliaSonera litigated a data-retention case to the Swedish Supreme Court after it was ordered to release the names and addresses of the operators of Swetorrents, a torrent site in Sweden, or face a fine of SEK 750,000 (*Computer Weekly* 2009). The company claimed that customer privacy commitments trumped IPRED. The site ceased operations before the Swedish Supreme Court heard the case. An important legal question then and still at stake was exactly how IPRED creates a legitimate exception to the privacy principle in force in the EU (the Data Protection Directive; Directive 95/46/EC).

As with ACTA, the onset of buyer's remorse came as popular anti-IPRED bias suffused the blogosphere.

Bloggers and net activists established websites denouncing the implementation of IPRED, and created other sites to keep track of the anticipated court cases that followed from implementation, and petitions started in opposition to the law. Moreover, the youth sections of the political parties unified themselves in their struggle against the implementation of IPRED. Cryptography experts raised the issue that a more widely anonymous Internet [from the rapid adoption of anonymizing services] would make it harder to find and counter other types of criminality, such as terrorism and child pornography. (Larsson and Svensson 2010, 7)

As privacy debates raged and people en masse began cloaking their identities online by using IPREDator software and other measures, it became clear that the EC had not clearly foreseen some of the consequences of beefed-up IP enforcement.

Patent law, which is ordinarily overlooked as a channel for colonization, also fell into the conflict over IPRED. Efforts to include stronger software-patent protections in IPRED failed. Those debates disclosed competing interests in the software industry and stimulated public concerns about the application of patent law to software. The organizations opposed to patent revisions benefited from strong support from open-source software groups and Sun Microsystems (FIPR 2005, para. 3). NGOs including the Foundation for a Free Information Infrastructure (FFII) convinced policy makers that stricter patent laws would disadvantage EU innovators competing with technology from the United States (EFF 2007; FIPR 2005). Opposition groups focused on the harm that strengthened software patents would do to the open-source community and pointed to potential economic damage (FFII 2005). The FFII noted, for instance, that Microsoft lobbied for stronger software-patent protections in the EU but lobbied for the opposite in the United States, suggesting that Microsoft wanted to create an unfair competitive edge over its European rivals (FIPR 2005, para. 3) in a form of policy arbitrage.

But overall, the IPRED negotiations demonstrated the preeminence of free market over nonmarket social values in the EC. The corporations supporting IPRED had more influence with rule-making bodies than their opponents, but the winner of any particular policy debate was usually the side that best articulated the trade and investment benefits accruing to Europe as a common market, rather than consistency with the EU's values supporting privacy and access. The social and cultural implications of the policy were concerns of secondary importance. Similarly, the primary obstacle to passage of IPRED2 for many months was legal uncertainty about the enforcement power of the EC rather than fears of the social and cultural harms the directive could cause in the EU (EFF 2007).

Had IPRED not passed the EP, IP would still have had strong protections across Europe. National implementations of IPRED duplicated and made redundant many national laws protecting IPR. And even with incomplete or uneven implementation of IPRED, "rights holders still have legal recourse through domestic laws" (Agarwal 2010, 799). For example, Sweden has the Swedish Patents Act (1993) and the Act on Copyright in Literary and Artistic Works (2005). Sweden's Copyright Law contains civil and criminal provisions similar to those of IPRED, with a maximum prison term of two years for "grave infringement" of copyrights, patents, and trademarks. Germany has legislation that addresses patents and copyright (1994) and the enforcement of IPR (2008) (Agarwal 2010, 799). Germany's Law on Copyright and Neighboring Rights provides an option for criminal prosecution and civil remedies, including monetary remuneration and injunctive relief. To the consternation of U.S.-based trade groups lobbying for uniform implementation of ever-stricter standards throughout Europe, Germany and Sweden "provide different sanctions for violations that are committed for personal, rather than commercial, gain" (800).

IPRED2

The justification for IPRED2 "stresses threats to national governments" such as organized crime (Open Rights Group 2011).[12] After its first reading in the EP, the measure moved to the Council of the European Union (it and the EP constitute the EU legislature) in 2007. In late 2010 the EC withdrew the proposed directive, leaving the EU without a formal rule for enforcing mandatory sanctions for the new copyright crimes.

IPRED2 proposed a rule for the kinds of mandatory sanctions provided by IPRED as originally drafted. It directed member states to pursue criminal penalties against purported copyright violators, including "everyday consumers" trading files privately and criminal organizations (Agarwal 2010, 813). The criminalization of file sharing was supported by transnational and multinational media corporations and some software corporations (FIPR 2005, section 1, para. 4). Supporters claimed that maximally strict IPR enforcement was needed to offset the economic problems caused by counterfeiting and infringement (ECDGT 2010, para. 2).

Critics pointed out that the proposed IPRED2 had been drafted "as a result of the EU's view that counterfeiting and piracy constituted a pressing international problem that was not being adequately addressed by current laws" (Agarwal 2010, 800). But one of the reasons for its being withdrawn was that the EU failed "to provide evidence as to how requiring Member States to implement uniform criminal measures [would] deter commercial infringement or mitigate its economic or social consequences" (Agarwal 2010, 813). Moreover, the proposed measures did not directly address the purported problem of counterfeiting: "IPRED2's current provisions fail to serve as a substantive response to the specific [counterfeiting] problem the EU initially recognized" (801). Perhaps even more problematically, IPRED2 used vague definitions of aiding, abetting, and incitement (791–792). Agarwal argues that amendments to either TRIPS or to IPRED itself would achieve the EU's goals without risking new human rights violations or threats to national sovereignty (813–814).

Stefan Larsson sees the IPRED and IPRED2 negotiations as evidence of a legitimacy crisis in the EU's Information Society program and overreach by copyright maximalists: "The fight against file sharing risks being drawn into legislative contexts of fundamentally different origin and legitimacy. A significant predicament . . . [is] that a directive that is drafted to fight terrorism, an activity with extremely low legitimacy in social norms, can end up in including the struggle against illegal file sharing of copyrighted content, an activity with extremely high legitimacy in social norms" (2011, 28). The "predicament" seems to repeat in history, with terrorism and child pornography having served for years as useful pretexts for regulationist communication policies worldwide, policies that were later extended to criminalize other activities.

Data Retention Directive

As discussed in chapter 1, pirate politics targets suspicionless surveillance policies for abolition, among others. Freedom Not Fear rallies were held in European cities every year from 2005 to 2013 in opposition to the EU's Data Retention Directive (Directive 2006/24/EC) and called for its repeal.[13] The directive, adopted in 2006, harmonizes laws requiring telecoms and ISPs to preserve personal information and telephone and Internet activity logs for up to two years, for "the purpose of the investigation, detection and prosecution of serious crime, as defined by each Member State in its national law" (article 1, section 1). The directive is oriented toward the prosecution of "serious crime," although there is no definition or prior European legal standard for the term.

Originally introduced after the Madrid and London bombings in 2004 and 2005, the Data Retention Directive reversed a former directive requiring telecom and ISP user logs to be erased as soon as they became unnecessary for billing (S. Larsson 2011, 20). Privacy-monitoring bodies within EU institutions roundly criticized the reversal as an illegal and unjust infringement of privacy (20). Swedish media surmised correctly that the Data Retention Directive would quickly be employed for copyright enforcement, since the "Data Retention Directive may aid the IPRED and the copyright holder's case against illegal file sharing" (21). Sweden's Riksdag delayed implementation of the directive, arguing that it duplicated Swedish law. The legislature was consequently fined and referred to the European Court of Justice (Humeau 2011). Sweden finally passed the directive in 2012, over objections from the Pirate and Green parties (Deutsche Welle 2012).

Telecommunications Reforms Package and ACTA

Besides IPRED2 and the software-patent provision of IPRED, two more significant harmonization measures failed to pass the EP in part because of resistance based on concerns about damage to the cultural environment online. The EU Telecoms Reforms Package (2007/0247 [COD]), which advanced through the EP from 2007 to 2009, would have increased the burdens on telecoms and ISPs of copyright policing (S. Larsson 2011). With both Pirate MEPs campaigning against the measure, a "last-gasp push" by activists finally derailed the entire Telecoms Reforms Package (Pignal 2009). The breaking point came with a proposal by Christian Engström to formally establish Internet access as an inalienable human right, which would have required a court order in order to restrict a person's access to the Internet (Vandystadt 2009b). The so-called Proposal 138 of the Better Regulation Directive and Proposal 166 of the Citizen Rights Directive of the package generated intense debate "in media, on blogs, and in the EU Parliament and the Council" (S. Larsson 2011, 23). These proposals forbade restrictions on users' access to the Internet "in any way that infringes their fundamental rights, and . . . that any sanctions should be proportionate . . . [and] require a court order" (23). Proposal 138 presaged the declaration in 2011 by the United Nations that Internet access is a basic human right and that disconnection by the state is unlawful under international law (LaRue 2011).

Finally, ACTA, an agreement to create a supranational body for global IP enforcement, was signed by the Council of the European Union in 2011.[14] The Council adopted ACTA during the winter holidays in "a completely unrelated meeting on agriculture and fisheries" (Sutton 2011). By 2012, ACTA had been ratified by the United States, Australia, Canada, Japan, Morocco, New Zealand, Singapore, and South Korea. ACTA, which is WTO compatible, requires signatory countries to promote and coordinate corporate efforts to suppress copyright and trademark infringement. As with the other regulations reviewed here, ISPs are the primary target of ACTA. The law limits their liability exemptions as mere conduits (S. Larsson 2011, 24). It was feared that ACTA "could lead to 'voluntary agreements' by Internet intermediaries to restrict Internet access and to monitor and censor Internet communications under threat of legislation or criminal sanctions" (Sutton 2011). Activists envisioned a secretly negotiated, final draft of ACTA that would provide for involuntary disconnection if it were ordered by the legal department of a media company.

Provisions for involuntary disconnection were purportedly removed from the text before publication of the USTR's 2011 draft, because of disagreements about judicial review. Yet the draft retained provisions that can restrict Internet access to prevent "imminent violation" of IP laws (Sutton 2011). And leaked discussion papers regarding ACTA's "three strikes" plan instructed ISPs, after being informed of a potential copyright violation by a third-party copyright holder, to send warning letters to users accused of copyright infringement (Ryan and Heinl 2010, 3). After two of those letters, if the alleged infringer was accused again, he or she would in some capacity be banned from Internet use. In its harshest form, a three strikes law could ban someone from using the Internet over mere accusations of copyright infringement, without judicial oversight, operating on the supposition that individuals sharing information online are guilty until proved innocent of IP theft (Braman 2007, 179–180; FFII 2010). Critics surmise that "ACTA . . . shows that there are strong international forces that seek to extend the means of enforcing copyright undemocratically, at the expense of ISP neutrality" (S. Larsson 2011, 24).

The potential consequences of those provisions did not sit well with many MEPs, who were actively fielding new waves of post-IPRED digital rights protests. The WikiLeaks diplomatic cable leak that uncovered "three strikes" provisions in ACTA in 2010 revealed strong Swedish and Italian resistance to ACTA, a secretive process of negotiations by member states, and evidence of "member states' explicit intention to avoid collaborating with international organizations on intellectual property, thereby circumventing opposition from developing countries that have questioned the need for additional IP enforcement agreements" (Sutton 2011). Before the final rejection of ACTA by the EP in 2012, Brazil and the Netherlands declined to sign the plurilateral agreement, and Switzerland and Mexico hedged, the legitimacy of the trade fulcrum being openly discussed by negotiators.

The fulcrum's value to the common market comes in exerting commodification leverage under the application of force. The force distorts communicative rationality by excluding entire categories of society as stakeholders. Of all their critiques, opponents of ACTA such as the FFII and the various pirate parties reserved one argument in particular to challenge the agreement's legitimacy: public interest and civil society groups were locked out of secret negotiations. Requests for access to ACTA-related documents were blocked in the United States after President Barack Obama claimed a state-secrets privilege over the negotiations. Eventually, pressure from the EP spurred the public release of an ACTA draft (EP 2009; Love 2009). Sunshine penetrated the secret ACTA negotiations for the first time after Knowledge Ecology International (KEI) obtained, via a Freedom of Information

Act request, a list of negotiating participants from the USTR (Love 2009). To no one's surprise, the list included representatives of Sony Pictures, Time Warner, the MPAA, and the RIAA. Corporate lobbyists such as the IIPA were also consulted. The sunshine revealed that, as with the EUCD, IPRED, and IPRED2, those key stakeholders, as direct consultants to the USTR in high-level negotiations, enjoyed far greater participation in decision making than the civil society groups that opposed ACTA. The negotiator list given to KEI included only one representative organization that supports cyberliberties, Public Knowledge.

The ACTA protests across European cities in early 2012 visibly demonstrated solidarity and popular support for cyberliberties. After the EP voted 478–39 to reject ACTA on July 4, 2012, KEI published dozens of public statements. KEI director James Love wrote, "There is real democracy in Europe, and a real parliament. And wow, what a social movement for fundamental rights! ACTA could never overcome its deficit in legitimacy and perspective. The US Congress has to take a deep breath and reign in USTR and other rogue anti-democratic agencies that are treating the public like enemies of the state" (KEI 2012). Former Green MEP David Hammerstein wrote:

Today in Strasbourg a massive majority of the European Parliament voted to reject ACTA. It is the first time the European Parliament has rejected an international treaty already signed by the European Commission (and by 22 of 27 EU member states). The power of the EU's legislative branch has been clearly reinforced as it has not been a rubber-stamp for questionable EU trade proposals as it has been in the past. Of even greater importance, European civil society has emerged as a very powerful actor that can no longer be dispatched by EU institutions with the traditional "participate a little, then we'll decide with our industry buddies." (KEI 2012)

Even before the EP rejected ACTA, opposition to the agreement from the pirate parties and digital-rights groups was joined

by India, China, and other developing nations, which claimed
that ACTA violates TRIPS and imposes unfair enforcement costs
that would be better spent on infrastructure and public health
(Geist 2010). Those objections, which resembled ones made during previous trade-related disputes, illustrate lingering concerns
about cultural imperialism (Mosco 2009, 73).

Click, Click

The upward IP ratchet is unlikely to pause. In another instance
of forum shifting in search of a more congenial site for operating
the IP ratchet, the United States moved its focus from the lagging
Doha Round of WTO negotiations to another secretive plurilateral agreement, the Trans-Pacific Partnership Agreement (TPP),
which "contains a chapter on IP enforcement that would have
state signatories adopt even more restrictive copyright measures
than ACTA" (Sutton 2011). The TPP's partnership countries are
Australia, Brunei, Chile, Malaysia, New Zealand, Peru, Singapore,
and Vietnam. The IP negotiating group was added in the third
negotiating session in Brunei in 2010 (Barfield 2011, 1). Pirate
parties in Canada and Australia made early calls for the rejection
of the TPP by those countries' parliaments.

The TPP moved the ratchet when Susan Schwab, USTR for the
outgoing George W. Bush administration, preemptively "forced
the TPP onto the trade agenda by throwing a trade policy dart to
the incoming Obama administration" (Barfield 2011, 1). Specu-
lation abounds "whether and how far the TPP will move beyond
the existing WTO rules" in TRIPS, and whether the ratchet will
lever up extra notches by producing a so-called WTO-plus agreement (5) that would finally strengthen software patents after
IPRED2 died. The government of New Zealand argues that "overly
strong protection beyond TRIPS could impede innovation in

developed countries and would harm economic development in developing TPP countries" (5). Other countries have reportedly joined New Zealand in pressing for health exemptions to any stronger patent protections.

The struggles over cyberliberties since the EUCD suggest that pirate politics' success in winning new rights for Internet users and defending remaining rights depends upon access to current and relevant information about EC decision making, the presentation of both economic and normative justifications, engagement at both EP and national levels, the cultivation of inter-industry alliances, exploitation of the boomerang effect to apply pressure from global civil society on legislators, and what Bennett and Segerberg call "connective action" (2012). From the EUCD to the Telecoms Reform Package to ACTA to the TPP, these initiatives, taken together, illustrate the path dependence of European copyright law, which has been modified without respect for the qualitative changes in technologies of information sharing since analog reproduction (S. Larsson 2011). Path dependence, in this sense, means that new IP initiatives rework and retrace prior reforms, even if the initiatives do not, in the end, move the IP ratchet up a single notch. Even with some serious setbacks to the maximalist trade agenda, the cumulative momentum of these initiatives may be said to have helped set the "facticity" (Braman 2006, 325) of maximalism in the mindset of regulators and citizens. But the momentum provides as well a greater basis for backlash against Internet colonization.

The Role of WikiLeaks in Shaping Information Policy

Daniel Domscheit-Berg (2010), formerly with WikiLeaks and recently with OpenLeaks, explains that WikiLeaks favored

Sweden, Belgium, and the United States as locations for their servers and operations because of the countries' relatively strong free-speech protections for publishers and journalists. Thanks to WikiLeaks, we now know that, besides the normative lag among Internet users in member states, some European states themselves have resisted the formal implementation of some EU-EC directives, finding them to be at odds with national and regional laws that provide greater access to knowledge or greater protections for personal privacy. WikiLeaks revealed that it is partly because of this resistance that the content industries have leaned even harder on Sweden and other governments for regulationist reforms. WikiLeaks also demonstrably enabled the SPP to adjust and adapt to ACTA developments in the absence of access to other information.

On his blog, Falkvinge interprets the leaked U.S. embassy cables as showing that the IIPA and PhRMA's joint lobbying efforts at the EU led to EU "lapdoggery."

Since 2006, the Pirate Party has claimed that traffic data retention (trafikdatalagring), the expansion of police powers (polismetodutredningen), the law proposal that attempted to introduce Three Strikes (Renforsutredningen), the political trial against and persecution of The Pirate Bay, the new rights for the copyright industry to get subscriber data from ISPs (IPRED)—a power that even the Police don't have—and the general wiretapping law (FRA-lagen) all have been part of a greater whole, a whole controlled by American interests. (Falkvinge 2011)

Falkvinge finds support in the cables to suggest that the EU aligns closely with the United States on trade treaties handling IPR and services, even when doing so goes against its own interests.

The quarter million WikiLeaks cables, which are classified as "sensitive" by the U.S. embassy in Sweden, outline, among many other things, cumulative efforts by the post on behalf of the IIPA and PhRMA to influence Swedish politics and law for the benefit of those trade groups. The embassy appears to have coordinated messaging in the midst of contentious public debates surrounding The Pirate Bay trial and impending votes on copyright reforms.

For example, the embassy pulled back from the brink of an openly hostile trade relationship and set some new boundaries for the trade relationship with Sweden in Cable "Stockholm 09-141." The so-called Silverman cable recommends that Sweden not be added to the Special 301 list, because of perceptions of positive developments by the chargé d'affaires, Robert Silverman: "Post continues to engage very constructively with the [government of Sweden], and has good access and a good working relationship with key senior and working level GOS officials. The actions taken since last year's review strengthen the legislative framework and provide better enforcement tools for combating piracy" (U.S. Department of State 2009b, para. 2). Despite its conciliatory tone, the Silverman cable presents a checklist of demands made of the government of Sweden by U.S. officials. The checklist, which contains both completed and hoped-for items, provides the most direct evidence yet of U.S. influence on Swedish information policy:

1. Adopt the copyright law amendments on injunctive relief against ISPs and a "right of information" to permit rights holders to obtain the identity of suspected infringers from ISPs in civil cases

2. Prosecute to the fullest extent the owners of The Pirate Bay

3. Increase the prosecutorial and police manpower devoted to criminal Internet piracy enforcement

4. Commence a national criminal enforcement campaign to target source piracy and large-scale Internet pirates

5. Ensure that rights holders may pursue the new civil remedies easily and quickly

6. Take an active role fostering ISP-rights holder discussions to effectively prevent protected content from being distributed without authorization over the Internet. (Falkvinge 2011)

Rickard Falkvinge's close reading of the WikiLeaks cables permitted the SPP to develop its maturing mission and objectives. Without the cables, the Pirates would have little or no information about the negotiations. I summarize the checklist here to illustrate how Falkvinge sees it as a smoking gun, demonstrating a greater need for transparency and openness in negotiating IPR treaties.

Point 1 refers to the adoption of three strikes, data retention, and ISP-disclosure rules. The Silverman cable showed that Sweden was poised to exceed the identity-disclosure requirements of IPRED, given contemporaneous negotiations by the Renfors Commission, a Ministry of Justice unit tasked with developing procedures for handling file-sharing cases. The SPP critiqued the Renfors Commission's "lopsided" proposal to adopt three strikes without judicial review, which Falkvinge calls "extrajudicial censorship" (Falkvinge 2011).

Regarding point 2, Falkvinge notes that the executive branch is prohibited by law from interfering with the judiciary, and signals that the U.S. State Department could stand accused of corrupting the Swedish justice system by pressuring the executive branch to pressure the judicial branch in turn. Falkvinge interprets point 3 of the checklist as meaning, "Transfer scarce police resources from investigating real crimes and devote them to safeguarding American monopolistic interests against ordinary citizens." He notes that a Swedish task force had to be established in 2011 to handle the instant backlog of lawsuits filed by Antipiratbyrån and the IFPI, which included cases against a fifteen-year-old who downloaded movies to a school computer and a sixty-year-old who shared music files noncommercially.

On point 4, and related to the education discussion of the previous chapter, Falkvinge excoriates the promulgation of pro-IP "educational" material sent to Swedish grade schools as being "politically biased, only tell[ing] half the story, or . . . directly and factually wrong." Regarding point 5, Falkvinge relates that in late 2011, the Swedish minister of justice, Beatrice Ask, announced a new traffic-data-retention rule meant to shift the enforcement focus from organized heavy crime to petty-fine crimes, including file sharing. Falkvinge reads point 6 as meaning that ISPs would lose immunity for carrying infringing content in some circumstances.

Besides Falkvinge's discursus on the cable dump, the dump itself enabled activists to develop near-time analyses, critiques, and responses to information policy that was being formulated behind closed doors and kept secret. It confirmed the suspicions of many that ACTA had legitimacy problems, and did so at a time when decision makers were avoiding consultations with delegations critical of the plan (Sutton 2011) and also withholding publication or publicity about their work for as long as possible (LQN 2011). WikiLeaks cables confirmed the strong role of the U.S. embassy in Sweden in motivating The Pirate Bay raids in 2006 and the subsequent prosecutions, and in pressuring Sweden behind the scenes to accept the enforcement and "educational" provisions in IPRED (Masnick 2010b; TorrentFreak 2010b).

WikiLeaks became enmeshed in pirate politics in ways that enhanced public deliberation about information policy and information age politics. The revelation of the U.S. embassy's checklist, especially, enabled the SPP to reaffirm its agenda and campaigns as relevant and useful. The cable dump and its contents in many ways validated the SPP's campaigns for increased openness and transparency in implementing the various

European directives, by unveiling problematic information policy that had been allowed to develop in secret. The appearance of the Silverman cable highlighted the shared dedication of the SPP and Julian Assange's WikiLeaks to expose injustices or corruption kept hidden by complicit, official means.

Remaining User Rights under European Law

The failures of the Telecoms Package, ACTA, and IPRED2 almost certainly forestalled further erosion of the communicative rationality of the European Internet, by suppressing the IP ratchet for a period. They also bought time for pirate politics to campaign for defense of the lifeworld online. This section offers a brief survey of user rights under the evolving regulationism. They are weak, but not entirely missing.

The E-Privacy Directive (Directive 2002/58) codifies antispam and opt-in provisions, along with restrictions on the use of cookies. Opt-in and cookie regulations provide a modicum of consumer privacy protection not available in the United States. The differences have created trade friction and may have slowed the expansion of some online media and advertising firms into European markets. But online, the right to personal privacy stops at IPRED. The E-Privacy Directive provides a good example of the EU way of managing the single market; EC regulations anticipate the consumer's expectations, often at the expense of the citizen's interests (Gollmitzer 2008).[15]

With software becoming a commercial service, and personal archives moving to cloud computing, it is notable that the EU affords software users valuable rights. A Council directive (91/250) from 1991 provides software users and researchers rights unavailable in the United States. These include "the right to make a back-up copy; the right to observe, test, or study a computer program; the right to test for error correction; and the right to 'decompile' or reverse engineer" (Dreyfuss 2004, 34). The European Database Directive (Directive 96/9/EC) "includes a right to perform acts inherent to normal usage and the right to re-utilize substantive parts of a database" (34).[16] Both directives predate the takeoff phase of broadband Internet in Europe. In subsequent trade negotiations, owners' rights rather than users' rights have set the legislative agenda, so pirate politics finds possibilities for ongoing campaigns for software users' rights.

HADOPI (France) and the Digital Economy Act of 2010 (UK)

Although the specific emphasis of this book rests on Sweden in the context of broader EU-EC reforms, and Germany features a strong Pirate Party presence also, two more countries' information politics deserve mention. France and the UK were among the early and eager adopters of the new wave of IP authoritarianism. They moved vigorously to advance the IP ratchet and, in so doing, generated sufficient opposition to launch pirate parties of their own in 2009.

The Creation and Internet Law was France's answer to a three strikes or graduated-response system (Ryan and Heinl 2010). It is commonly referred to as the HADOPI law in reference to the acronym of the legislative body—the Higher Authority for the Distribution of Works and the Protection of Copyright on the Internet (Haute Autorité pour la Diffusion des Oeuvres et la Protection des Droits sur Internet)—that the law creates. The law, which came at the conclusion of talks between the entertainment industry and ISPs with the French government, gave HADOPI the authority to oversee complaints from copyright holders over infringement (Senate of France 2009). HADOPI has the authority to suspend individuals' Internet use for up to a year for copyright infringement and can enforce software-blocking measures at institutions where infringement is found to be committed. Initially, HADOPI was to be given the authority to punish infringers on its own, but the law was rewritten to include judicial review (Ryan and Heinl 2010).

HADOPI turns infringers in to the courts after three warnings, and, in addition to forfeiting their Internet access, infringers may be fined up to €300,000 and face up to two years in prison (Ryan and Heinl 2010). A private investigative company for the copyright industry (Trident Media Guard) tracks down infringers and turns them in to HADOPI. The HADOPI law also contains specific statutes modifying the education system. Article 15 of the law amends the education code for students in mandatory music and art classes, from grade school through college, to be "informed of the dangers of downloading and of illicitly making available works or objects protected by a right of authorship or a related right for artistic creation" (Senate of France 2009). Remarkably, the law mandates that secondary education classes dealing with informatics and Internet certification be taught by teachers with specialized backgrounds in IP law (Senate of France 2009, article 16). As in Sweden and elsewhere, these pro-IP changes to the education system are clear examples of the constitutive power of information policy (Braman 2006, 19).

The UK's Digital Economy Act of 2010 received criticism both for its IP authoritarianism and for being born obsolete. The act does not introduce a pure three strikes system, but establishes a foundation for one by giving the Office of Communications, the UK communications regulator, the ability to eventually instate such a rule if the measures that the act introduces prove ineffective (BIS 2010). The Digital Economy Act's three strikes provisions are to come into effect in March 2014, and expose libraries, hotels, and bars with open WiFi to infringement claims (Sweeny 2012).

ISPs, again, bear the responsibility for gatekeeping. The act requires ISPs to track suspected copyright infringers and keep a record of how many times infringement is committed. Copyright holders can at any time apply for a court order to get the names and addresses of alleged infringers and take legal action (BIS 2010). Then, copyright holders send cease and desist letters to the alleged infringers. Should that measure fail, copyright holders can take accused infringers to court for criminal litigation (Ryan and Heinl 2010). ISPs are charged with the responsibility of devising methods for tracking copyright infringement and are required to slow the connection speeds of suspected infringers, block their Internet Protocol addresses, and suspend their service (Digital Economy Act of 2010, sec. 9).[17]

The act was fiercely debated in the United Kingdom. British Pirates claimed that the negotiations had been unfairly influenced by the British entertainment industry and Hollywood studios. Among the first effects was the block placed on TPB's Internet Protocol address by several ISPs, including British Telecom (BT). The British Pirates sounded the alarm.

BT became the latest major UK ISP to block its customers from accessing The Pirate Bay yesterday evening. BT has also blocked the additional addresses added by The Pirate Bay in response to the block in recent weeks. Within minutes of the block being put in place, The Pirate Bay made additional addresses available to circumvent the block. [Party leader] Loz Kaye made the following statement on BT's block: "Last year, Vince Cable promised the country that the site blocking provisions of the Digital Economy Act would not be implemented, this was widely interpreted as meaning the coalition is opposed to web censorship. Blocks on Pirate

Bay have effectively short circuited the democratic process. . . . Our Internet policy is not being run by our elected representatives, it is being dictated by the music industry." (PP-UK 2012)

In classic hacktivist fashion, the Pirate Party-UK (PP-UK) set up a proxy service for TPB (https://www.pirateparty.org.uk/Proxy). Although it makes no explicit alliances with WikiLeaks in its web communications, the PP-UK supported Julian Assange publicly when he sought refuge in the UK in 2012.

Developing Alternative Models for Information Policy

TPB, like Christiania, persists as an active enterprise in spite of Europeanization, projecting an alternative collective project into a normalized social landscape. Even after its new purchase and trusteeship in 2012, as a territorial space, Christiania seems forever imperiled by colonization. TPB forestalls a virtual collapse; it revirtualizes itself by distributing a "pirate bay dump" of the site's source code and index that fits on a USB drive, while its cocreators continue to develop disruptive technologies.

Information policy is among the most culturally sensitive domains of the law, since it mediates the emergence of new forms of communicative action, and since it "affects the nature of facticity, meaning the ways in which data treated as 'facts' are created, perceived, and incorporated into decision-making" (Braman 2006, 325). It is also critical for political and economic systems. Leaving aside the implications for civil liberties and an open society, failed information policy in an information society can damage the structures and processes of communication and even stifle economic growth (Benkler 2006, 135). Pirate politics helped Europe avoid some worst-case scenarios at the EU and EC. Work on the EP's Committee of Legal Affairs by Pirate MEP

Christian Engström, in particular, moderated the worst abuses of cyberliberties lurking in draft legislation (see Engström 2009b for examples). In addition, Engström disclosed the existence of an "IP observatory" for rapid IP law enforcement strikes (Masnick 2010a), a discovery that could lead to new campaigns against abusive new information policy.

As the struggles over the IPRED software-patent expansion, the Telecoms Reform Package, and ACTA suggest, some strategic uses of communication frames can be more persuasive than others. The framing of information policy issues by Benkler presents the basic proposition that the pirate parties advance: sharing and creating content not only can contribute value to a global information economy, but also is the only way that the economy can be sustained. The challenge for reformers is not to convince policy makers to reject the notion of information as a commodity, but to emphasize examples of the intrinsic benefits of information as a shared resource (Benkler 2006). For instance, free software may be especially beneficial to small businesses, and the chilling effects of the criminalization of copyright infringement may weaken Europe's $25 billion content-creation sector (PAC 2009, 117) by stifling innovation (FIPR 2005, section 1, para. 4). Indefinite copyright-term extensions may be shown to harm competition, and data retention policies may prove to be too expensive for the courts, police, and ISPs to maintain. And so on. The networking that led to the rise of public awareness of the new ACTA commitments is a blueprint for how the intellectual commons movement can participate in future policy debates (EP 2009). Pirate politics has the potential to address broader policy areas than file sharing and surveillance, including software, "nonlinear" and video-on-demand digital cable, Internet Protocol television, e-books, games, and web publishing.

This chapter has enumerated the critical components of the EU Celestial Jukebox, highlighted the influence of pirate activism on its development, explained the strategic importance of WikiLeaks for pirate politics, tracked the remaining user rights in the InfoSoc regime, and proposed how pirate politics might get even more traction and expand its horizon. Although the political history of the Pirate Party is still young, and springs from ongoing dynamics with TPB, the SPP can be said to have already exerted a moderating influence on the InfoSoc-EUCD template.

The moderation is significant, given the path dependence of European copyright (S. Larsson 2011) and the compelling logic of the single market for digital culture. The ratchet and fulcrum, together, lever network power into new social contexts that are detached from the lifeworld, all the while replicating the system's bureaucratic-administrative logic. Seen this way, Europeanization combines easily with spatialization achieved through the combined force of the 301 process, forum switching, and shift to plurilateralism. Information policy is captured by private industry groups such as ACTN and IIPA seeking full-spectrum controls over the terms of access to knowledge, and control over its policing. A risk of these outcomes is that information policy proposals offering greater communicative rationality never get a proper hearing. Pirate politics reveals a legitimacy deficit in information policy, and in so doing, reveals the existence of a transnational, networked public prepared for confrontation with the system over the terms of access to the information society.

3   Technoculture versus Big Brother

The transformation of Swedish copyright and privacy rights into part of the economic-administrative complex of the EU-EC system sparked protests that led to mobilization over data retention and IPRED. The conflicts indicate that while juridification of the online lifeworld is occurring rapidly, it is still incomplete. The Internet has not yet become entirely toll taking, like a car park, a movie theater, or the iTunes store. Everyday anonymity is attainable online, although active surveillance makes deep anonymity improbable. Darknets flourish as never before, even as users' online lives are tied increasingly to shopping carts and virtual checkouts, software as a service, expiring e-books, non-interoperable software, innumerable log-ins, and bulging cookie caches. As industry builds new silicon and software cages, some users become bolder in avoiding them and in asking about the need for new rights, such as the right to be forgotten (Rosen 2012).[1]

Pirate politics is intended to "ensure that Europe chooses a better road into the information society" (Engström, qtd. in Woldt 2009). The colonization of cyberspace discursively thematizes and politicizes "once unquestioned and apolitical assumptions and practices by drawing them into the administrative domain," in turn creating "questions where once there were apolitical assumptions" (Crossley 2003, 296). The salient assumptions concern the adoption and spread of IP authoritarianism in the EU information society, a development that gave rise to pirate politics. Europeanization here is a "stirring up of cultural affairs that are taken for granted" and "furthers the politicization of areas of life previously assigned to the private sphere" (Habermas 1988, 72). The Pirates test how well the political and legal systems are able to address new political grievances. Since the Pirates participate in two national political systems and the EP, they break down some barriers that previously relegated most NSMs to permanent outsider status and redraw the system boundaries.

Pirate politics arose amid pervasive anxieties about the EU's path to the information society (Spender 2009). The anxieties are clearly expressed in the legal and social studies literature about loss of privacy online, the lock-down of digital culture in copy protection, the criminalization of sharing practices, and anonymous and often unknowable negotiations over global information policy (Gillespie 2009; Johns 2010; Peñalver and Katyal 2010; Zittrain 2008). The anxieties cluster around blocked access to information, knowledge, and power, as well as around violations of expectancies about using the Internet autonomously. As Klaus Eder claims, such fears of relative deprivation do not concern subsistence and survival, or the loss of elite privilege. They are middle-class fears. "The cultural basis upon which new social movements were built" is a specifically middle-class culture in pursuit of the good life and lifestyles that support "personal aggrandizement, autonomy, and competition" (Eder 1995, 38).

Charting the Waters of Cultural Politics

Pirate politics claims—or perhaps invents a claim—that because of the colonization of the Internet, users have been excluded from the social means of realizing their identities.[2] Party leaders express the claim as a moral crusade to restore a lost online agora, *integritet* (*privacy* in Swedish), and enhanced personal freedoms. Its universalist goals, conflicted as they are, have internationalized the appeal of pirate parties, and they offer a template and an umbrella movement for cyberliberties activism.

With some exceptions, (see, for example, Calhoun 1993), sociologists tend to identify NSMs with postindustrialism and to consider their emergence historically as having coincided with a globalized and high-tech economy characterized by networking and connectivity (Beck 1992; Kelly 2001, 30; Offe 1985). So it is appropriate that an NSM that is "about" media and communications and is reflexively oriented toward reforming cultural and social codes bears a family resemblance to reflexive NSMs of earlier generations, especially ecology and feminism. According to Alberto Melucci:

In the current period, society's capacity to intervene in the production of meaning extends to those areas which previously escaped control and regulation: areas of self-definition, emotional relationships, sexuality and "biological" needs. At the same time, there is a parallel demand from below for control over the conditions of personal existence [that are] . . . part of society's self-reflexiveness in information and communication production and sociability. (1989, 45–46)

The popular demand for control over the conditions of personal existence is even stronger in an age of "networked publics" (Varnelis 2008). The heightened visibility of pirate politics since the takedown of TPB, its registering in the popular imagination of the technoculture, and its transformation from a protest movement into a formal political party require fuller sociological explanations. The sociology of countercultural movements, NSMs, and "anti-political politics" (Havel 1991) takes varied forms, but typically highlights civil society as a constitutive field built from the articulation of universal rather than particular interests. Movements articulate these interests in communication directed to society at large, rather than to the state, in expressive campaigns that make use of alternative and radical media.

Here we encounter another discrepancy between pirate politics and inherited NSM theory. The institutionalization of pirate politics could suggest that NSM theory has had only limited purchase on the movement since it became a minority party in Sweden. The NSM theorists Claus Offe (1985), Jean Cohen (1982, 1985), and Nancy Fraser (1985) explicitly exclude from consideration groups that organize into political parties. The transformation of pirate politics may have demonstrated a type of social learning that reduces path dependence on the IPR ratchet by fighting it and producing a social actor that is, as Christine Kelly puts it, "uniquely capable of reflecting critically on [its] context" (2001, 110). It is still the case, however, that civil society sustains pirate politics, even as it operates in the margins of the political system. The Pirate Parties of Sweden and Germany, and many of their siblings, together with varieties of collective action by WikiLeaks, Telecomix, and Anonymous, exhibit classically "defensive" postures in symbolic politics directed to society at large. Melucci (1989) calls social movements "nomads of the present" for the way that they move and adapt to colonization, highlighting disjunctures and disruptions along the way.

Pirate politics is marked by a collective identity and transnational cultural formations engaged in conflictual symbolic politics performed for networked publics. This approach to activism and networking considers cultural movements and political movements together as social agents that present multiple "spheres of publics" (Calhoun 1995) in global civil society with alternative visions of modernity. Pirate politics emphasizes personal autonomy, lifestyle concerns, solidarity and collective identity, and shared political grievances; like other cultural movements that NSM theory considers, pirate politics mobilizes in fluid, submerged, and informal networks (Buechler 1995, 442). Offe describes movement politics as the period in which "multitudes of individuals become collective actors [in] highly informal, ad hoc, discontinuous, context-sensitive, and egalitarian modes" (1985, 829).[3] In its networking, identity politics, and periodicity, pirate politics shares affinities with contemporary antiglobalization movements and advocacy networks (Crossley 2003; Della Porta and Tarrow 2005; Keck and Sikkink 1998), environmentalisms old and new (Boyle 2008), and even earlier forms of Western cultural politics from the nineteenth century (Calhoun 1993).

Although Habermas's contribution to NSM theory has been "problematically tied to a structural strain approach to movement mobilization" (G. Edwards 2008, 300), the theory of communicative action still provides an excellent framework with which to support the basic proposition that lifeworld defense in cyberspace is oppositional and based in an NSM. It can show that the structural changes required by what has been called digital capitalism (Schiller 2003), informational capitalism (Braman 2006), technocapitalism (Kellner 1989), or cybernetic capitalism (Mosco 2009) tend to impoverish communicative rationality, unless countervailing social movements target colonization. I agree with Crossley (2003) that colonization is occurring through the market and neoliberal economics as much as (or even more than) through welfare statism, and that Habermas's colonization thesis is an artifact of historical reflection upon the 1960s and 1970s made in the 1980s, before the rise of the New Right, the imposition of its new economic model of neoliberalism, and the dismantling of the welfare state. For critical media studies, it is the creation and growth of the Celestial Jukebox, together with its technical and juridical infrastructures, that deserves critique.

NSMs and Communicative Rationality

With their radically participatory forms of collective action and lifeworld defense, NSMs are communicatively rational, which is to say that they are "oriented to achieving, sustaining and reviewing consensus—and indeed a consensus that rests on the intersubjective recognition of criticizable validity claims" (Habermas 1984, 17). Communicative rationality is both the result of social learning processes (Crossley 2003, 292) and a contribution to learning.

The approach pursued here takes late modernity to represent the working conditions and social environment within which pirate politics operates. The late, or "new," modernity (Kelly 2001) approach to NSM theory tries to explain the continuity of new political forms and processes with those of the past, and disputes the claims of postmodern theorists that information-age capitalism is qualitatively different from industrial capitalism. Late modernity affords NSMs opportunities to achieve consensus on social problems arising from colonization.

Habermas, Offe, Beck, Eder, and Cohen and Arato, on the one hand, for example, analyze contemporary forms of politics as coping mechanisms for the strains of a hyperrationalized and superdifferentiated culture and society. Touraine, Melucci, and

Laclau and Mouffe, on the other hand, assume that NSMs demonstrate a definitive break with the politics of capitalist modernization. In late modernity, basic rules of sociality and power are reformatted completely, and information in the aggregate replaces social subjectivity as a source of agentic change.

The discontinuity thesis is less well suited to explain the persistence of capital accumulation, authoritarianism, and technocratic power in the transition to the information society, and—apart from new social movements—does not discern a basis in social action for social change. There is a lack of evidence to support the contention that cultural logics have mutated to the point that late modernity breaks with the past, and considerable evidence to the contrary (Habermas 1987b). As mentioned previously, adherents of the new modernity approach to NSM theory, with its empirical focus on civil society, argue that NSMs by definition cannot formalize into political parties or formally engage the state, but must dwell "elsewhere" (Kelly 2001, 38). Otherwise, NSMs are part of the political subsystem. That position rejects the possibility that social movements can be both formal and informal, developing in both the system and civil society. It also loses sight of the fact that, as Melucci (1989) points out, movements group and select elites. While "not *easily* adaptable to the existing channels of participation and to traditional forms of political organization, such as political parties," new forms of conflict produce political actors ready to take advantage of political opportunity structures, such as those described in chapter 2 (41; emphasis added). Here Melucci seems to be opening the door for an argument based on continuity. Moreover, the networked public sphere developed after NSM theorists began to demarcate system-lifeworld boundaries defined by organizational communication, complicating the picture greatly. Therefore, pirate politics may be considered a maturing, even already matured, NSM, with its grappling hooks planted firmly in the system.

Some skeptical objections to the NSM approach deserve to be addressed early. Havel's famous antipolitical politics offered as a description of NSMs has been interpreted to mean that they evade a political calculus altogether. For example, Russell J. Dalton and his coauthors (Dalton, Kuechler, and Bürkli 1990) argue that NSM theory violates rational choice models in political science, which emphasize instrumental, means-ends reason and games. This objection misses the mark, however, since NSM theory looks instead for evidence of communicative rationality.

Kevin Hetherington (1998), in making a strong argument that emphasizes the "neotribal" dimension of protest movements, offers nearly the opposite proposition. He considers NSM theories to be overly dependent upon subject-centered reason and the imputation of means-ends rationality and instrumental thinking to postmodern agents. Key to his argument is the claim, following Raymond Williams, that movements express a "romantic structure of feeling" and not rational discourse and debate (83). Hetherington's analysis of neotribes and their affective messages comes very close to expressing what NSM theorists themselves see as one of cultural movements' main characteristics, which is the use of affective rhetoric and moral reasoning in their appeals. Yet the criticism of NSM theory for mistaking feeling for reason focuses not on the effects of communicative action, but only on its symbolic contents. NSM theory considers the content of communicative rationality to be agnostic and to admit all kinds of claims, even "romantic" (Hetherington 1998) or "mimetic" ones (Kelly 2001, 21). What is more important is the communicative context—how and where those claims are posed, and the effects of the communicative action. It is indeed the case, as Kevin McDonald argues, that "the sensuous is at the center of the movement experience" (2006, 95), because movements navigate among the structures of the cultural lifeworld. It is the analyst's role to ascertain a movement's communicative effects.

Communicative Rationality and Social Learning

The lifeworld's defining characteristic is its communicative rationality (G. Edwards 2004, 116). Freedom underpins the notion of communicative rationality as well as the notion of free culture, which is the rallying cry and source of meaning for pirate politics. The persistence of communicative rationality in cyberspace makes it possible for cyberspace to become a field of contention where people demand new rights and freedoms. In an anthropological study, Ulrika Sjöberg (1999) describes Swedish teenagers' attraction to online sociality as a "free zone" for developing their own relationships and identities largely independently of parents and other authority figures. Other ethnographic work shows that hacker culture, which is oriented both to the freedom to tinker and to cyberliberties generally, is embedded in the lifeworld online, where it flourishes (Coleman 2010).

The communicative rationality expressed in NSMs is implicit in their critique of growth of the system (Habermas 1981, 34). Gemma Edwards argues that 1980s-era movements could "reassert communicatively rational action against the distortions of an increasingly intrusive system" (G. Edwards 2004, 117), and asks why this should not also be true for movements of the twenty-first century, especially the anticorporate movement. The present work, which is informed by the work of Nick Crossley (2003), contends that pirate politics, like the anticorporate and anti-globalization movements, has inherited the reflex of 1980s-era movements against "the tendency in advanced capitalist societies for the lifeworld to be engulfed by the growing 'economic-administrative complex'" (G. Edwards 2004, 115) of state and economy. The communicative ties to family, work, leisure, education, and health, for example, are beset by "state bureaucracy, legal regulation, political socialization and economic privatization" (115) as consequences of system growth. The basic logics of system growth, which are juridification and commodification (115; Habermas 1987a), result in a loss of "both freedom and meaning" (Crossley 2003, 294) and in "privatized hopes for self-actualization and self-determination" (Sitton 1998, 78). Like the antiglobalization movement, the pirate movement can be viewed as "a response to or manifestation of the growing crisis of neo-liberalism" (Crossley 2003, 299).

Besides being a domain of freedom and a basis for critique of system growth and crisis, communicative rationality offers a model for social learning through social movements. Identifying social goals and guiding a general conversation about those goals can lead to broader, more negotiable validity claims. Communicative rationality is counterposed to instrumental rationality and system contexts, and is considered incommensurable with them, although communicative and instrumental rationality frequently coexist and commingle in the same social contexts. Dietz explains the distinction for Habermas's theory:

Central to Habermas' philosophy is the distinction between strategic and communicative action. When involved in *strategic action,* the participants strive after their own private goals. In doing so they may either compete or cooperate, depending on whether their goals oppose each other or rather coincide. When they cooperate, they only are motivated *empirically to* do so: they try to maximize their own profit or minimize their own losses. When involved in *communicative action,* the participants are oriented towards mutual agreement. The motivation for cooperation therefore is not empirical but *rational:* people respond e.g. to requests because they presuppose that these requests can be justified. The basic condition for communicative action is that the participants achieve a common definition of the situation in which they find themselves. This consensus is reached by negotiations about the validity claims raised. (Dietz 1991, 239; emphasis in original)

In everyday speech acts, validity claims are presumed—and can be challenged—as claims to truth, justice, and sincerity. Speech acts regularly succeed, with all three validity claims accepted. When a claim is challenged for any of these reasons, negotiations and an accounting process can begin to save the speech act from failing. "Only those speech acts to which the speaker assigns criticizable validity claims do motivate the hearer on their own to accept the speech act offer, and only because of this foundation do they become the mechanism for effective coordination of action" (Dietz 1991, 240).

The notion of communicative rationality, which inherited these key features from George Herbert Mead's (1962) pragmatic tradition, relies also on the notion of speech acts from John Searle in order to show that language coordinates social action through activities such as requests and promises. Request fulfillment is a foundational form of social coordination (Dietz 1991, 236).

The universal pragmatics Habermas develops from Mead and Searle is further elaborated with the aid of Émile Durkheim, who is read as providing an evolutionary development from mechanical to organic solidarity (Habermas, in Dews 1986, 104). The key point of this sociology of communicative action is that communicative rationality evolves through processes of social learning.

Mead and Durkheim offer a developmental basis for explaining "the increasing reflexive fluidity of world-views, . . . a continuing process of individuation, and . . . the emergence of a universalistic moral and legal system" (111–112). These are processes that can be accelerated by NSMs in lifeworlds that are reproduced in communicative action and carry emancipatory potentials.

Habermas himself does not pursue communicative rationality to cyberspace and, indeed, seems to associate networks and networking with functional integration common to the system (2001, 82). Extending the lifeworld concept to cyberspace, "where people interact face-to-device with each other in conditions of telecopresence," yields a basis for finding intersubjective communication, or "we relationships," online (Zhao 2004, 92). The Internet's open and transparent communication protocols, near ubiquity in developed markets, and normative acceptance into personal life has led researchers to discover many ways in which cyberspace has become interwoven with the communicative lifeworld that Habermas describes. For example, Christopher Kelty (2008) describes the development of "recursive publics" in the lifeworld online, and Gabriella Coleman (2004) describes the egalitarian sociality of hackers who participate in the FOSS movement. In an influential study, Kevin McDonald considers NSM communication in general, and antiglobalization protests in particular, to exhibit "fluidarity" and networking dynamics in which "forms of action flow from one network to another, in a process amplified by the Internet and communications networks such as Indymedia in the US or the Tactical Media Crew in Italy" (2002, 44). NSMs use fluid communication networks in cyberspace as a means of building solidarity around a critique of system growth.

These characteristics suggest that twenty-first-century NSMs can, in fact, support communicative rationality, help decolonize the lifeworld, and replace juridified routines with social learning. They cannot rely on ideology to manipulate supporters, but instead use deliberation to promote noninstrumentalist and nonmarket social relations in existing social institutions. Ecology does so with respect to the environment as a socially constructed legal and physical domain; feminism does so with respect to families, the workplace, and education, among many other social institutions; and cyberliberties do so with respect to information policy, education, and the state.

Hacking the Software of State

The entry of pirate politicians into the EP served as a reminder to voters and MEPs that that the original vision of the information society was technocratic and oriented to system growth at the expense of the lifeworld. Establishing the basic rights of Internet users, especially when challenged by prosecutors of IPR, ISPs, and the state, was indefinitely postponed in favor of prioritizing trade-related IP claims of owners and purported owners.[4] Pirate politics offered a corrective.

NSMs are good decolonizers because communication underlies the informational mode of production that spawned them: "*Information resources* are at the center of collective conflicts emerging in highly differentiated societies" (Melucci 1995, 113; emphasis in original). Sandra Braman (2004) too makes this argument, noting that conflicts over information characterize the inner dynamics of the information society. Seen from the perspective of communicative rationality, pirate politics participates at multiple levels in protecting the lifeworld online from colonization. It intervenes in colonization processes and transforms them into renegotiated politics. It disrupts and reprograms expectations about communicative rationality in the Internet generations. It expresses conflicts over codification, specifically, the cultural and legal codes that order law and policy regimes for communication and information. The conflicts are spread over overlapping social codes, cultural codes, software code, and legal code (Lessig 1999; Melucci 1989) and are visible as hacktivism, culture jamming, radical media, and pirate politics. For Melucci (1989), the purpose of movement politics is to give notice that there is a fundamental clash between cultural codes and technocratic routines and to discover and enjoy social interaction with affinity groups. Pirate politics supports Melucci's basic proposition, but amends it to include greater participation in information-society decision making.

The recursive characteristics of pirate politics are strengthened by virtue of its constitution in geek culture and cyberspace. Kelty's work illuminates geek culture as an affinity-based solidarity network built on collaborative work and production: "Geeks imagine their social existence and relations as much through technical practices (hacking, networking, and code writing) as through discursive argument (rights, identities, and relations). In addition, they consider a 'right to tinker' a form of free speech that takes the form of creating, implementing, modifying, or using specific kinds of software (especially Free Software) rather than verbal discourse" (2005, 214). Geeks' creative work often includes speech acts manifested as code and technical writing, but they present to one another at numerous conferences, or "cons," hackerspaces, and at other convivial places as well. And although geek culture is collaborative, it favors meritocracy, tournaments, contests, debates, and deliberations as ways to drive the creative process and generate innovations.

More broadly still, the ascendancy and internal organization of pirate politics are owed in part to the connectivity of its origins in the technoculture. The network provides the underlying structure for association, identity formation, and political mobilization. In their mobilization, pirates help us see how civil society and its public sphere have become much more like "a network for communicating information and points of view" (Lim and Kann 2008, 78), with the Internet providing "a convivial milieu in which various political uses are thriving and new tools for political criticism and commentary are emerging" (80). Moreover, since NSMs operate in the margins of system and lifeworld, at the edges of the public sphere, they communicate the edges of the network, together with the shifting frontier between system and lifeworld, to the political system itself.

Online deliberations can coordinate social action with considerable egalitarianism and legitimacy. Asynchronous communication tools help manage many-to-many, large-scale discussions and deliberations. Tools such as deliberative polling, online Citizens' Juries (which appeared in the UK in the 1990s), Unchat, and e-liberate (facilitating something like an online version of Robert's Rules of Order) facilitate A2K, deliberation, and debate (Lim and Kann 2008, 83–85). U.S. examples of online deliberations include AmericaSpeaks 21st Century Town Meetings and Wisdom Councils facilitated with groupware (85). During and after such deliberations, the "Internet's inherent conviviality enables rapid, widespread mobilization" (90) among marginal actors.

The Swedish and German Pirate Parties are famous for using commercial services, including Twitter, Facebook, and PasteBin, to communicate among their members and followers. But, being geeks, they have developed their own solutions to the practical problems of social coordination. The board members of the Pirate Party of the state of Berlin worked with the Public Software

Group in Berlin to develop LiquidFeedback, FOSS for supporting democratic deliberation and agenda-setting among members of a political party or any other organization: "With the platform, issues that would previously only gradually find their way to the national leadership through local, district and state organizations can quickly gain momentum and importance, so that they can then be voted on at party conferences" (Becker 2012).

The software platform uses a flexible set of rules to promote direct democracy.

> The basic idea: a voter can delegate his vote to a trustee (technically a transitive proxy). The vote can be further delegated to the proxy's proxy thus building a network of trust. All delegations can be done, altered and revoked by topic. I myself vote in environmental questions, Anne represents me in foreign affairs, Mike represents me in all other areas but I can change my mind at any time.
>
> Anyone can select his own way ranging from pure democracy on the one hand to representative democracy on the other. Basically one participates in what one is interested in but for all other areas gives their vote to somebody acting in their interest. Obviously one may make a bad choice once in a while but they can change their mind at any time. (Nitsche n.d.)

The linguistics professor Martin Haase holds no elected office, but has the largest number of delegated votes on the German Pirates' LiquidFeedback system, making him at one time "the most powerful member of Germany's burgeoning Pirate Party" (Becker 2012).

LiquidFeedback enables the Berlin (state) Pirate Party to address a broad range of concerns quickly and to expand its platform beyond IPR issues. At the time of this writing, other areas included matters of "personal freedom," including a basic income guarantee for artists, FOSS developers, other creators, students, and the unemployed; personalized learning goals for students (a goal that requires rejection of the EU's Bologna Process); and the "right to get stoned" (Urbach, personal communication). The use of networking and voting software to develop those priorities in real time is of great practical and symbolic significance for the German Pirate Party. It permits rapid adaptation of the party's agenda to political changes. The cultural director for the Berlin state Pirate Party has expressed hope for replacing the entire German parliament with LiquidFeedback and promoting a "platform neutral state" with respect to technology policy (Urbach, personal communication).

Besides software development, other examples of pirate mashups include the politicization of information embargoes in the War on Terror by WikiLeaks' data dumps of classified U.S. materials (such as the "Collateral Murder" footage); the cooperation of the SPP in providing WikiLeaks with Internet connectivity after it was disconnected forcibly; and the ongoing SPP campaigns against further Europeanizing reforms to copyright, information, and communication policies. Repertoires of contention built up from "technologies of dissent" (Curran and Gibson 2012) include hacktivism, working on FOSS, mobilizing transnational protests, assisting rebel communications during the Arab Spring revolts, and blowing the cover off hegemonic power cloaked in bureaucratic secrecy.

Supporting activities aside, the ultimate objective of pirate politics is pragmatic: the reform of legal code governing information in the EU, where a "growing constitutional legitimacy gap" requires bringing ever more parties to the bargaining table when law and policy are being decided (Marsden 2011). In addition, the pirates' primary goal provides a fresh opportunity for social learning, since the rejection of the European Constitution in 2005 set the stage for political parties and pressure groups to reframe rights as specific policy initiatives for national and European parliamentary campaigns. The codification of new digital rights might mark the growth of communicative rationality and help legitimate the European system of law (see chapter 4). Habermas characterizes law as "a medium through which . . . structures of mutual recognition . . . can be transmitted . . . to the complex and increasingly anonymous spheres of a functionally differentiated society" (Habermas 1996, 318). In its deliberative mode, pirate politics emphasizes widespread participation and open processes in the shaping of information policy, which could lead to legitimacy, since "the ultimate grounding of operative law [is] in critically self-conscious, discursive processes directed to universal reasonable agreement" (Michaelman 1996, 311).

The EC, rather than the EP, seems to be a domain where pirate politics can have the greatest impact on the legitimacy of information policy. The EC forecloses many of the communicative spaces for "debate, analysis, and criticism of the political order" (Fossum and Schlesinger 2007, 1), as was seen in the IPRED and ACTA negotiations. The institutional insulation of commissioners from national demands is at the root of criticisms of its technocratic, distanced, and unaccountable operations. Andy Smith proposes three reasons for this: they are likely to be upstaged in the media by MEPs and national ministers; the multiplicity of conflicting roles the commissioners play detracts from their ability to "speak with one voice"; and they have weak connections to journalists in national media (2007, 228). The division of the commissioners' labors is not defined nationally, partly for fear of claims regarding unjust proportionality of distribution, and so calls for commissioners to be accountable to their member states frequently go unanswered. The legitimacy of the European commissioners' representativeness "is constantly undermined by the specific set of rules and expectations that lie at the heart of the institutionalized role of a European Commissioner" (228). That the commissioners themselves are generally not information technology experts or even literate in new media contributes to their vulnerability.

MEPs have greater direct responsibility for their national constituents, but that responsibility is shared with coalition partners. Elected to their positions from national parties running in national referenda in member states, MEPs are also members of European party groups. Pressure from European party groups exerted on individual members—through committee assignments, resource allocation, and other methods—helps set agendas, shape MEPs into reliable and disciplined party members, and (sometimes) assist them in withstanding countervailing pressure from home (Hix, Noury, and Roland 2006, 496). MEPs' committee work, such as Pirate MEP Amelia Andersdotter's work with the Committee on Industry, Research, and Energy and the Committee on International Trade, is an appropriate place to look for ongoing influences on decision making.

Eder envisions the EU-EC as a learning game for the entire region. The EU creates a never-ending iteration of transnational mechanisms linking communicative spaces where legitimate claims to authority can be made, contested, and accepted in a performative fashion (Eder 2007). The EU has structured an "open organizational field" (59) that generates "interorganizational games" (56) and builds "a polity on a set of normative rules that reproduces itself through the continuous addition of further procedural rules" (58). From a shared normative basis, deliberation can proceed through the EC and other "elite publics" that engage in "collective learning" among multiple groups and constituencies, including national groups (58–59). So in theory, pirate politics could propagate social learning throughout the EU-EC and help forestall impending legitimacy crises.

The NSM Sociology of Pirate Politics

Social movements are like an autoimmune response to lifeworld colonization. They rely on cultural, pretheoretical knowledge, and intuitive know-how to make their defense. I argue that pirate politics, as a variety of cultural politics, is a symptomatic response to technocratic incursions by the law, public policy, and private corporations into formerly undisturbed areas of the lifeworld. It is partly institutionalized, in a conventional, parliamentary sense, and partly fluid. Both NSM and resource mobilization (RM) approaches to NSMs conceive of movements as alternative forms of sociality that superannuate participation in traditional party politics and emanate exclusively from civil society. RM approaches address organizational strategies, and NSM approaches address identity-maintaining strategies; there are also hybrid methods. Jean Cohen (1985) finds the two approaches mutually informing, while Gemma Edwards (G. Edwards 2008) finds the RM approach to be implied in Habermas's critical systems theory. As I have argued elsewhere (Burkart 2010), accepting both perspectives provides for the adoption of complementary system-and-lifeworld viewpoints that cover empirical domains accessible through systems functionalism and *Verstehen* sociology. What is important here is that pirate politics operates in civil society with formal connections to the political system.

RM looks for what Charles Tilly calls "repertoires of contention," or "clumps of performances that describe the forms of claim-making available to any particular set of political actors, including the government, and the likely consequences of making such claims" (2006, 22). RM reveals that the pirates are, by and large, "rooted cosmopolitans" (Della Porta and Tarrow 2005) who participate in global flows and exchanges of ideas, media, and software. Their local stances on issues of concern are informed by transnational networks of activists. Their disputes over the legal disposition of communication resources such as sharing platforms and software, and their critiques of anticompetitive business practices in media and software, appeal especially to students, technical professionals, software engineers, and entrepreneurs, whose freedom of movement in the European common market often expands their horizons beyond the scope of their localities.

RM is possible in the transnational collective action of pirate politics in part because of the "complex internationalism" exhibited by European states, their international institutions, and nonstate actors (Tarrow and Della Porta 2005, 231). Cyberspace has become a critical resource for mobilizing an observable, international repertoire of contention since the Zapatista uprising in Mexico (Castells 2001; Terranova 2001). NSM theory can offer insights into the symbolic identities of the rooted cosmopolitans. Culturally, pirate ideology partakes of the New Communalism, a postmaterialist sensibility that has been present in countercultural movements since the 1960s (Turner 2006). New Communalists look for their own technical solutions and for local alternatives to dependency on big business and big government. The spirit of the Whole Earth 'Lectronic Link of the 1980s still animates this "maker" aesthetic, which serves as an ethic as well.

As noted previously, pirate politics is a middle-class phenomenon. While pirate politics may avoid explicit class appeals in its rhetoric, appears to have denatured class characteristics, and has an ecological understanding of culture, its orientation to the pursuit of economic growth through formal politics exhibits the material interests of the middle and aspiring middle classes. Playing politics in a complex international setting, especially, commits the players to a conservative aim of "preserving a familiar world in which politics intervenes as the executive arm of supposed economic progress," as Habermas says in a critique of the Stuttgart 21 protests (Habermas 2010). (Stuttgart 21 is a proposed railway and urban development project.)

Like other social movements, pirate politics shows the tension between a form of communalism aspiring to achieve a creative commons such as the public domain, and the kind of strong individualism consistent with anarchistic ideals. The communitarian perspective valorizes the public-good characteristics of digital cultures and the natural (that is, untrammeled) Internet as providing an online agora. The libertarian-anarchistic perspective prizes autonomy, hacker competencies, and personal privacy. Both perspectives fit comfortably within the culture of the new middle class. The SPP's leaders make individualistic appeals to a shared ethic of personal responsibility and, in place of policing and regulation, call on users to self-regulate and demonstrate personal "integrity" (Troberg, in O'Dwyer 2009; to follow). As noted previously, the party leadership has aligned the SPP with the leftist Greens, although Falkvinge was a member of the conservative Moderate Party's youth wing (Phillips 2009) and even describes himself as "ultra-capitalist" (Tuccille 2009).

The pirates have a liberal precedent in engineering history. Engineers have made calls to action in defense of ethical values and visions of a better social life, based on their knowledge, understanding, and work with technology. Engineering societies in the early twentieth-century United States provided a precedent for an ethic of engineering professionalism and class solidarity (Layton 1986). Around the time of World War I, members of the engineering professions "seemed to be searching for a redefinition of their place in society, to be seeking a more active role in resolving the problems (both technical and social) of the day. Many of them seem to have concluded that engineers, by reason of their training, experience, and social position, could develop a different, and superior kind of leadership than that exercised by business" (Meiksins 1988, 219).

The "revolt of the engineers" (Layton 1986) against the corporate businesses that hired them presented a wide variety of grievances and calls to action, but none with unanimity or even a plurality of support. Activists issued calls for greater social responsibility of business, greater unity among those working in engineering professions, more humane treatment at work, and a meritocracy for professionals (Meiksins 1988, 222). But the profession was split between probusiness elites, also known as the "patrician reformers," and activist bench workers, whose unions and professional organizations ultimately failed to unify around the incipient ethic of engineering professionalism. Pirate politics is marked not by professional societies and unions issuing calls to action, but by far looser affiliations of professional and preprofessional students, independent contractors, entrepreneurs, staff technicians and engineers, and geeky prosumers.

More implicitly, but perhaps just as importantly, pirates reveal the blocked ascendancy of technically skilled people into the middle and upper-middle classes.[5] Eder contends that the new middle class of knowledge workers is a technological and cultural elite, because of its members' greater conceptual ability, knowledge, and opportunity to become involved in politics than some of the working classes in manual industry (Eder 1995, 44n17). The long economic crisis of 2008 broke many of the popular assumptions about social mobility and deprived many knowledge workers of job positions that weren't "peripheral" or "decommodified" (Offe 1985). In this context, the pirate movement seems even more modestly middle class and limited. Blocked ascendancy underlies grievances voiced by people engaged in collective action with Occupy Wall Street, the "movement of the squares" across the Middle East, the Wisconsin protests, and the Spanish Indignados (Kennedy 2011). Ray Corrigan uses work and education examples to comment on the emotional appeal of SPP campaigns against ACTA: "If you start knocking people off the Internet for allegedly infringing copyright, those numbers [of Pirate Party voters] start to grow into the thousands, or tens of thousands, very quickly. It has a direct impact on their children's education and some people may need the Internet for their job. When IP has a real impact on lives that is when you start to see a backlash" (qtd. in C. Edwards 2009). Marina Weisband, the political director for the Pirate Party of Germany, considers pirate politics, the Occupy movement, and the Arab Spring uprisings to share a common antiauthoritarian orientation and a cosmopolitan understanding of networking as a political activity, in addition to their middle-class appeal (personal communication). Blocked ascendancy has meant, then, that the material concerns of the financial downturn starting in 2008 are increasingly being paired with the quality-of-life concerns of 2006 that launched the institutionalization of pirate politics.

The information policy objectives of the Pirate Parties of Germany and Sweden pose risks to some entrenched interests— including the entertainment industry, pharmaceutical and software companies, and private and public surveillance agencies.

Pirates do not advocate for the abolition, socialization, or nationalization of those businesses. The industries fear that broad implementation of the Pirates' platform would threaten the growth of their IP monopolies. That fear is, as yet, unfounded, since the SPP's veto powers on committees and its narrow focus area promote moderation of only the most radical of the system's proposals, such as the three-strikes rule. Pirate politics may be counterhegemonic and countercultural, but it is not antisystemic.[6]

The communicative position of the pirates discloses the congruence of class interests among students, independent software engineers, web developers, start-up entrepreneurs, and others whose livelihood and aspirations depend upon access to knowledge, sharing software and tools, and communicative rationality online and in creative workplaces. Control over the norms that regulate the workplace, and resistance to colonization there, necessarily involves struggle (G. Edwards 2004, 127). But the struggle exhibits "a self-understanding that abandons revolutionary dreams in favor of radical reform that is not necessarily and primarily oriented to the state. We shall label as 'self-limiting radicalism' projects for the defense and democratization of civil society that accept structural differentiation and acknowledge the integrity of political and economic systems" (Cohen and Arato 1992, 493).

The pirates' self-limiting characteristics accommodate certain forms of domination while resisting others. As previously mentioned, the pirates recognize that a transnational corporate elite opposes their interests systematically. An antitransnational corporate attitude therefore pervades pirate politics, which, like antiglobalization politics, emphasizes local and independent capital and entrepreneurship and opposes complex internationalism. The massive antiglobalization demonstrations in Seattle,

Prague, and Geneva targeted international governmental agencies (principally the WTO and IMF) for their unaccountability to the people they affect and for their service to transnational corporations. The main message of the protests was that globalization creates "an unjustifiable legitimation deficit" (Crossley 2003, 298). The protests are joined by business projects such as Fair Trade and by culture-jamming operations such as Adbusters, which reject and problematize the role of the consumer and refocus attention on the producers of mass consumer culture. These filial movements decolonize the lifeworld by addressing the "moral grammar" of society and reclaim a moral organization for society from the economy (298).

Damage and Repair across the Generations

Habermas emphasizes the "disenchanted, internally differentiated and pluralized lifeworlds" (Habermas 1996, 26) that are inhabited by participants in social movements, whereas many second-generation critical theorists who have engaged with Habermas on NSM theory emphasize the reparative work of the movements' communicative rationality. Jean Cohen and Andrew Arato interpret NSM politics as having transformative potential, for three reasons. First, the emergence of cultural modernity, and of differentiated spheres of science, art and morality, organized around their internal validity claims, "carries with it a potential for increased self-reflection (and decentered subjectivity) regarding all dimensions of action and world relations" (Cohen and Arato 1992, 524–525). Second, modernity selectively favors processes of system integration as it institutionalizes its rational potentials, leading to the colonization of the lifeworld. Third, there is a "two-sided character of the institutions of the contemporary lifeworld, that is, the idea that societal rationalization has entailed institutional developments in civil society involving not only domination but also the basis for emancipation" (524–525). Cohen and Arato identify NSM politics with the rejuvenation of civil society and the public sphere, focusing on the "institutional potentials" of the "shared cultural field" (511).

For Axel Honneth (1979) too, communicative rationality can be found in the margins of the detached system and lifeworld. He argues that communicative action there "should generate both a meaningful self-definition and a critique of domination through a process of collective social criticism which would reach the social space where theoretical enlightenment can be politically organized" (59). Working outside the second-generation critical theory tradition, but in some ways that are consistent with it, Alberto Melucci (1989) and Ron Eyerman and Andrew Jamison (Eyerman and Jamison 1998) argue that NSMs are regenerative and liberating forms of social transformation, and that the alternative forms of sociality and individualization that they offer are ends in themselves.

These perspectives share an interest in explaining that social movements can decolonize the lifeworld, but the process of decolonization remains unclear. In their platforms, Pirates seek to block some of the commodifying processes underlying the basic structures of technocapitalism and reinforced by the legal, financial, and political systems spanning the EU. Most important, pirate politics seeks to short-circuit the commodification logic of the Celestial Jukebox by blocking global DRM and restricting the kind of unlimited data retention that can make a universal panopticon out of customer relationship management or social networking.

Pirates target other colonization mechanisms as well. Linguistic colonizers include legal codes such as indefinite copyright renewals, European-wide patents for software and business methods, and the single-market initiatives covered in chapter 2. Those codes become integrated with nonlinguistic steering media, including the "network power" (Sassen 1999) of members of the IIAA and their industrial allies. Network power propagates colonization mechanisms across networks and code bases, including the DRM and automated surveillance and data-retention systems required to enforce three-strikes laws and DNS blocking, as well as the royalty rents such measures are designed to protect. In addition, Pirates target the secrecy and silence surrounding opaque treaty negotiations, which typically exclude public deliberations. The formal banning of DRM, or decriminalization of work-arounds to circumvent it, is a decolonizing Pirate objective that would ameliorate the conditions of unfreedom that obtain in the online "digital enclosures" (Andrejevic 2002) that are maintained by contract law, content management systems, and technical copy protections.

The Green-Pirate Continuum

Eder (1985) analyzes ecological counterculture movements as varieties of collective learning that can expressed as high levels of communicative rationality, particularly as they began to make distinctions between ecological and environmentalist ideologies, and between appropriate and inappropriate technology practices. There are important complementarities between the Greens and pirate movement, beyond the fact that Christian Engström joined the Greens/European Free Alliance in the EP.[7] Like an ecological counterculture movement, pirate politics is both a pressure group and a moral crusade (Eder 1988, 159). Both ecological crisis and the tragedy of the cultural commons share exploitation as a common referent. NSM theory permits us to consider pirate politics as a variety of environmentalism and, in turn, to consider environmentalism as a mechanism for social learning in European society (Kousis and Eder 2001, 26).

James Boyle claims that "precisely because we are in the information age, we need a movement—akin to the environmental movement—to preserve the public domain" (2008, xv). For Boyle, who writes within the juridical discourses of positive law, no social movement or social actor has yet arrived to "save" the public domain from politics as usual and crisis. Cultural environmentalism remains, for the most part, an acute awareness of the creative, technical, and legal resources needed to protect online culture. Its political sensibility treats life in the information age as a shared, communicative canvas, waiting to be painted upon with a mélange of tools and colors taken and remixed from the media environment: "The environmentalists helped us to see the world differently, to see that there was such a thing as 'the environment' rather than just my pond, your forest, his canal. We need to do the same thing in the information environment" (xv). Tim Jordan and Paul Taylor (Jordan and Taylor 2004) argue that, in fact, global piracy and hacktivism have communicated this same sort of understanding of the information environment for at least two decades.

As a moral crusade, environmentalism makes appeals to noneconomic social values that are programmed out of the public sphere by economic forces. Cultural environmentalism mimics environmentalism in its drive to transform the morals of society: "Environmental politics is action taken less on the environment than on society. . . . Environmental politics (including environmental policy) shapes not simply the environment: above all, it shapes society" (Kousis and Eder 2001, 7). In civil society, pirate politics takes society and free culture as its objects, whereas in parliament its object is the administrative-legal complex.

At the system level, the Greens' formation of policy for environmental regulations and the Pirates' participation in information-policy negotiations defend and preserve territories that are critical for survival. Both seek to develop better protections for commons perceived to be at risk of destruction. Because of the proximity of environmentalism to health and public safety concerns, it can unify public support around nearby areas of culture, even in political systems without as strong a political opportunity structure as Europe's. For example, Stephen Hilgartner compares the Pirates' information-policy agenda to the movement for consumer rights and workplace safety in the United States of the 1970s, and especially the call for "a voice in decision making in areas of environment hazard management and consumer safety" (2009, 197). Reformist social movements in the United States lobbied for and won the creation of the Environmental Protection Agency, the Consumer Products Safety Commission, and the Occupational Safety and Health Administration (197). In Europe, the Greens/European Free Alliance embraced a consumer rights agenda. The natural foods movement provides another point of comparison, since it communicates fears of the risks of "the impurity that is intensified by the technification of the industrial food culture" (Eder 1988, 159).

As Green parties spread across Western European countries in the 1980s, becoming formal players in coalitional governments, social scientists observed the development of an ideological continuum "from red to green" in radical politics (Fehér and Heller 1983). The exhaustion of labor-based political appeals for solidarity and lost political mandates of labor parties, together with the abdication of the revolutionary project by the New Left, coincided with the ascendancy of postmaterialist political ideologies. "Counterculture movements have reframed the modern relationship of man with nature as one of exploitation" (Eder 1996, 139), while labor exploitation was partly abandoned.

The appearance of nationally based pirate parties running for local, state, and national offices, while affiliating internationally, resembles the early organizing of the European Green parties and the Swedish Green Party (Miljopartiet de Grona). But no single crisis of the cultural environment, such as Chernobyl, has yet pushed the SPP into office. Following the Chernobyl disaster in the USSR and other environmental crises at home, the Swedish Green Environmental Party won its first seats at the Riksdag, running on a platform to shut down nuclear power plants, strongly enhance environmental regulations, and eliminate fossil fuel use in Sweden (Affigne 1990). The German Greens (Die Grunen) "emerged from sixties radicalism with an eclectic ideology and an aversion to 'bourgeois' politics," eventually developing a policy platform of ecology, grassroots democracy, social responsibility, and nonviolence (O'Neill 2000, 166). "The Green manifesto . . . developed the case for ecocapitalism, reconciling ecology with progressive social measures and radical economic policy. In effect, the Greens temper an ethical antipathy to capitalism by advocating a 'fair' employment regime, social justice, and better eco-management: namely, reduced work-time, job-sharing, eco-friendly job creation and retraining schemes" (171).

Part of the Greens' success as a political party in Germany is owed to its ideological transformation from fundamentalist and purist to pragmatic and accommodating. Environmentalism watered down ecology and even offered a development program that promoted economic growth in times of deindustrialization.

Through a combination of faith in technological progress, financial and technical participation by technocrats and elite knowledge workers, and political accommodation, the Greens provided a politically palatable program of energy savings, smart growth, and emission controls. The migration from social movement to political party demonstrated that whatever antisystemic dynamic the Greens once contained was rendered a merely reformist and moderate force for social change. Deep ecology and ecofeminism continue to resist colonization pressures in submerged networks.

New pirate parties are still being institutionalized from submerged networks. The Czech Pirate Party led the effort to "port" the Creative Commons framework for the Czech Republic and has participated in municipal elections (Carr 2011). The Pirate Party-UK jump-started the national conversation about the Digital Economy Act after the law received little attention during its passage. The Liberal Democrats placed repeal of the act's piracy provisions at the top of their list of 2011 convention items (B. Jones 2011). The Pirate Party of Canada, launched in 2009 as a blog, released a torrent tracker that tracks Creative Commons content; in addition, the party ran ten candidates for office in the 2011 elections (PPCA 2011). The Pirate Party of New Zealand held its first annual conference and policy-making meeting, operated a membership campaign, and registered with the New Zealand Electoral Commission in 2011 (PPNZ 2011). As ACTA faltered in the EP in mid-2012, the PPI continued to increase its membership, with the Russian Pirate Party officially registering in 2012.

Like European NSMs of prior generations, pirate politics cultivates materialist concerns related to survival and social mobility, and postmaterialist values focused on identity, expressive individualism, and "an authentic lifestyle where people interact as equals and as free persons" (Eder 1995, 33–39). Its agenda and communications contain decolonizing potentials that can serve as a heuristic for a European political system on the wrong path to the information society, and that in the process can contribute to a more robust European public sphere.

The cultural sensibilities of cyberliberties may endure, but if the Green parties of Europe serve as a guide, rapid institutionalization is likely to blunt the radicalism of pirate politics and underscore "the transitory nature of movement politics" (Offe 1990, 235). After all, the Pirates' success "has made politicians realize they have been complacent and need to catch up" on cyberliberties issues (Schultz, qtd. in Inskeep 2009). Identity politics in the movement, "a normal and perhaps even intrinsic part of a successful, democratic public sphere" (Calhoun 1995, 248), and ongoing, engaged participation will continue forming and revising the Pirates' political identities (273). Since the German Greens now claim to endorse the same technology program as the Pirates, it may well be the case that the Greens can become the "new pirates." And many Pirates would accept that outcome if it meant that the pirate agenda would advance (Weisband, personal communication). However, pirate activism is still performing a winnowing function for the political system by self-regulating for purity of ideological commitment. The rejection of ascendant figures such as Julia Schramm (contentious author of *Click Me* and formerly of the German Pirates) for bypassing pirate publication channels and selling out helps in the selection of new elites, and enables the movement to develop and practice its distinctive moral grammar.[8]

As the previous chapters have argued, pirate politics imposes political limits to commodification and juridification in ways similar to those used by other NSMs. In the negotiation of those

limits for ecology, a basic tension emerged between the ecological antigrowth imperative that makes the movement antisystemic, and smart growth or planned growth, a tension that
normalized the Greens as they moderated their positions over
the years. Pirates face the same normalization process, especially
regarding contradictions between individualism-communalism,
in-group—out-group distinctions between party and civil society activists, materialism and postmaterialism, and so on. Such
is the fate of the rooted cosmopolitans of the technoculture.

Meanwhile, cultural colonization, which is one effect of
Europeanization, materially diminishes the bases of support for
the social reproduction of the technoculture, taking its toll; for
example, the rendering of new martyrs from old heroes (Julian
Assange, Aaron Swartz, Gottfried Svartholm) extinguishes their
spark and serves notice for others. However, there is nothing a
political pirate likes any more than a fresh challenge from the
system, and new challenges provide political opportunities for
new campaigns. In calls for arousal or quiescence, pirate politics is constantly performing its identity politics for the approval
and support of networked publics.

Conclusion: The Legacy of Pirate Politics for Moral-Practical Reasoning

Pirate governance did not come from out of the blue. The assertion by many students and young pirates that the politicians are "declaring war against a whole generation" (Soares 2009) rests on an observable paradox: copyright criminality grows with the spread of the information society. The problem for critical media studies is to illustrate how the paradox is, in fact, not an unintentional artifact of information and media policy but, indeed, a carefully planned and coordinated outcome.

To address this research problem, *Pirate Politics* has provided a glimpse into one form of activism in global civil society that perceives the paradox to be a social problem with political solutions. In this context, pirate politics offers potentially transformative campaigns. I have described the pirate identity and lifestyle as countercultural, even though they are considerably less than radically alternative. Pirate politics shows clearly that communication media underlie transnational collective action around information policy and digital rights. I have described the double reflex of pirate politics, in which culturally embedded technological practices cultivate qualities of communicative rationality intended to codify digital rights.

The present case study provides updated tests of concepts designed to accommodate the appearance of horizontal and participatory processes of media activism that contribute to what Puppis (2010) describes broadly as media "governance." Pirate politics seems to support the claim that media governance is practiced by a range of interested constituencies besides official decision makers in the political system. In addition, the appearance of pirate politics supports the claims by NSM theorists that civil society remains a domain of growing importance for the study of politics and governance. Thus, my analysis included updated models for civil society and global civil society, including spheres of publics (Calhoun 1993), networked publics (Varnelis 2008), and complex transnationalism (Tarrow and della Porta 2005). These notions both update public-sphere democratic theory and illuminate pirate politics as a cultural phenomenon.

But since pirate politics often organizes and performs collective action in the porous and anonymous spaces of darknets, it becomes necessary to explore the role of such hybrid domains. Hackerspaces, Internet Relay Chat, and ephemeral message boards such as 4chan /b/ are semipublic fora with clubby in-group and out-group distinctions. So pirate politics operates in submerged networks while also making public claims for recognition and authoritativeness in global civil society. The organizational diversity of the players extends beyond the few cases focused on here (the SPP, the German Pirate Party, the IPP, and their precursor groups) to include, among many others, the Icelandic Modern Media Initiative; Julian Assange and Birgitta Jónsdóttir; Christopher "moot" Poole and Anonymous hacktivists; and coders, artists, and musicians. These agents collectively function as a countercultural movement under conditions of complex transnationalism.

Since complex transnationalism privileges juridical communication and tends toward lifeworld colonization, pirate politics is forced to operate counterhegemonically. It follows the money and the sources of system power, and attaches its grappling hooks to the system in the form of agents playing formal roles in political decision making while reporting on the policy-making process and encouraging mass resistance. By campaigning for and winning political office, developing alternative forms of sociality in formal systems, and blackmailing and hacking the organs of state, Pirates engage the political system directly, even as they continue to operate in unconventional pressure groups and loose, ad hoc networks. As counterhegemonic actors, the Pirates focus on power emanating from the EC and, especially, the United States. As Jonas Andersson notes, pirate politics "underscore[s] . . . the relational nature that many smaller European countries have towards the U.S.; being 'at the periphery of the center' entails an altogether rather specific experience in terms of a mix of awe, advantage, and resentment towards U.S. imperialist interests" (personal communication). For the SPP, this experience informs the critique of Swedish "lapdoggery" (Falkvinge) regarding the United States, the need to expose policy-making masquerades, and the assertion of new and alternative bases of law for greater political legitimacy.

Generally speaking, the Pirates' efforts are progressive in their orientation, and can be seen alongside those of other NSMs as "represent[ing] a non-reactionary, universalist critique of modernity and modernization by challenging institutionalized patterns of technical, economic, political, and cultural rationality without falling back upon idealized traditional institutions and arrangements such as the family, religious values, property, state authority, or the nation" (Offe 1990, 233). I emphasized the nonreactionary nature of pirate politics and argued that a vote for the Pirates symbolized a vote withheld from reactionary and right-wing competitors. But as previously mentioned, the institutionalization of the TPB was owed in large part to financial and technical assistance from Carl Lundström, a wealthy Swedish patron who has been reported to be a white supremacist and member of the anti-immigration group Keep Sweden Swedish (Expo Idag 2005). Lundström's 40 percent share of TPB earned him a fine and prison sentence after the infamous trial. After TPB came under fire for Lundström's role as its patron, a TPB spokesman, Tobias Andersson, said of Lundström's support that it was better that he gave the money to TPB than to Sverigedemokraterna (Lischka 2007), a xenophobic, nationalist, and ultraright-wing political party. It is noteworthy that Lundström did not participate in the SPP's formation or ascendancy, and self-described Pirates reject his racist or formerly racist ideas. But his early role in TPB raises important questions about the implicit links between pirate politics, populism, anti-Europeanization, and a romantic longing for lost origins.

Romanticism animates most utopian ideologies. Pirate politics exhibits some familiar aspects of cyberutopianism that also impel cyberliberties NGOs like the EFF, La Quadrature du Net, and KEI. An inherited feature of the 1960s-era New Communalism, cyberutopianism is a structure of feeling about the Internet that is romantic in its longing for an original design. It nurtures a heartfelt belief in the salvific powers of convivial networks, in spite of the growing facticity of copyright maximalism and policymaking for post-privacy. It is essentially optimistic and hopeful that the social agency of online communities can deliver the Internet from an impending crisis, and it is likewise sensitive to new threats. The theory of communicative action supports the underlying and positive dialectic of cyberutopianism and the teleology implicit in its model of communicative rationality.

There are two main reasons for qualifying this utopianism, and they address the Lundström concerns as well. First, as I argue, the model of social inclusion implicit in this utopianism is in fundamental tension with its middle-class identity and its political program for economic growth and development. Utopias cannot countenance second-class citizens. Second, pirate politics anticipates and addresses the "cyber-realist" critique of cyberutopianism and "Internet-centrism" (Morozov 2011). Because Internet centrists treat the Internet as a constant, they fail to take responsibility for defending their existing freedoms and for curbing the power of companies like Google and Facebook (335–339). But by working in the margins of civil society and the state, and within states, Pirates posit the Internet as a variable whose values are determined by politics, and then work the political system to reframe those values. The critique of cyberutopianism promotes the social shaping of technology over technological determinism, arguing for the case that reality is a "complex interplay between social and technological factors" and privileging neither set of factors (339). In fact, the Pirates' international mobilizations around IT and information-policy reforms are engaged and coherent, even oppositional and tactical. They demonstrate an intense awareness of—and participation in—the political, economic, and cultural shaping of technology. Moreover, the pirates distrust Europeanization as a cumulative consolidation and collection of executive powers, and they promote a restoration of power to the legislative branches of the EU and member states. They prefer promoting social learning to technocratic decision making. Above all, the pirates mean to break from the demonstrated path dependency of information policy by selecting horizontal strategies of governance in addition to traditional mechanisms of government, despite being a weaker player in governmental institutions.

New political realities could transform the basic Pirate platform. Because of their political realism and sustained engagement with state actors over the main tensions of the information age, the Pirates stand to accelerate a decomposition of the tripartite cyberliberties platform of free speech, privacy, and access that is often treated as a unitary set of "digital rights," such as I have described. This would be an ironic outcome were it to come to pass. But the powerful state actors against which the Pirates are aligned do indeed tend to undermine the purported unity of digital rights. The Julian Assange and Chelsea Manning sagas problematized those linkages poignantly by showing the willingness of states (at the time of this writing, the United States and the UK) to collude in removing free-speech protections from "new journalism" whistleblowers and to place explosive state secrets further out of reach of the public. The WikiLeaks case shows, among other things, that pirate politics drives an evolution of "reason of state."

Legal developments in the area of EU privacy protections—namely, the proposed "right to be forgotten" by social media sites—could additionally pit privacy against free speech and access. The right to access the unrevised historical record stands as a contemporary challenge to personal privacy rights and, especially, to a person's right to be forgotten. When a person's self-fashioning online requires deleting or editing out one's online history, must historians and journalists accept the intentional data loss? What about the state? Do corporations also enjoy the right to be forgotten? Pirates face the prospect of conceding some rights in order to defend others.

Then there are the theoretical arguments about a "post-privacy" condition by Christian Heller (2011) and the Pirate Julia Schramm. Post-privacy is the ontological condition created by the combination of data retention, surveillance, and the popularity of persistent identity services such as Facebook, Twitter, and Google. Post-privacy is a political-realist argument challenging the Pirates' rationales for enhancing privacy laws and opposing new surveillance measures. The argument raises legitimate questions. Is it delusional to fight for the right to personal privacy and to take active technical and legal measures to protect one's privacy? Would it be better to promote personal integrity and mutual respect for privacy online than to cling to technical and legal means of preserving personal privacy? Could it be more communicatively rational to act as if one's life is an open book? Does the post-privacy condition deprive pirate politics of a central purpose, or shatter the unity of digital rights? Assuming that pirate parties persist as formal organizations for the foreseeable future, the Pirate agenda may very well remain single-issue: a fight to preserve the unity of digital rights.

More broadly still, cultural environmentalism is under growing pressure to address the question whether to codify new rights at all. Ironically, codification of certain cyberliberties can become a self-limiting exercise of governance. Chapter 1 touched on the potential pitfalls of codifying old rights, such as the right to ramble in the countryside. The risks include any arbitrary diminution or dilution of traditional practices still legitimated by cultural norms. Seen as part of the system-lifeworld dialectic and Europeanization, the risks include replacing negotiated and substantive coordinating action in interpersonal relationships and democratic procedures with formalized routines, narrowly juridical thinking, and recourse to harms and damages as measured by existing tort law. With every impending click of the upward IP ratchet, Pirates face the choice of asserting a narrow positive right where none has been needed.

While codifying digital rights procedurally and formally seems like a communicatively rational course of action, it is possible to imagine how the broader pirate program could become hemmed in by settling for multiple, modest assertions of cyberliberties in the face of much more aggressive, ongoing colonization by the royalties-bearing industries and the state. For example, the codification of fair use under the DMCA could have restricted DRM much more broadly and clearly, but instead requires users to re-petition for specific exemptions from penalties for circumventing DRM. In another example, personal privacy protections from the ubiquitous closed-circuit television cameras in London could have been ensured by banning the cameras altogether, but instead have been restricted to blocking images appearing in the windows of buildings falling under the gaze of the cameras. On the other hand, the digital rights agenda could have come under even greater pressure if the pirate parties had been rebuffed entirely by law-and-order-seeking political parties. So it is now difficult to imagine a Pirate parliamentarian who would pass on the chance to make positive commitments to new rights and protections, however limited and self-restricting they might be. The Pirate MEPs have exercised real influence over the creation of new International Telecommunications Regulations of the International Telecommunications Union, the rejection of the ACTA, and the official condemnation of Hungary for authoritarian new media laws, for example.

Pirate Self-Fashioning

Since identity is symbolic, and since language underlies our symbols, language, gender, and sexuality shape the emergent identity of political pirates. In what I think is a happy feature of the Swedish language, there is a single word for both "integrity" and "privacy"—*integritet*—which may help inform the pirate identity in ways that unify the movement around a common conception of ethical action. Leif Dahlberg explains:

In Swedish legal language, *integritet* is short for *personlig integritet*, which translates as "privacy of the individual," typically denoting that only you yourself should have access to "personal" information about you, such as medical information, banking information, telephone records, and (of course) information about what websites you visit and what you do online. In other contexts, *integritet* in Swedish has a similar meaning to "moral or personal integrity" in English. That is, to have personal integrity is to have well thought through and consistent opinions that you don't change lightly (or to gain favor or money). It's another way of saying that somebody is a good person.

However, as is natural in (natural) languages, there may be contexts in which these two different uses (meanings) of the Swedish word *integritet* come into contact and start to communicate with each other. From a linguistic point of view, this is of course only bad language use, but from a literary/rhetorical point of view it may be quite effective. (Dahlberg, personal communication)

Whether or not it is bad legal Swedish to mix different contexts of the same word, *integritet* carries a heavier semantic load than "privacy" does in English. In the context of cyberliberties activism, the sense of personal integrity in *integritet* refers to personal autonomy and self-determination. It is easy to identify this use of the word in the publications of self-described Pirates (such as Anna Troberg, SPP leader). From another perspective, there appears to be a symbolic deficit in the English word "privacy" (in comparison with *integritet*) that is marked by a failure to recognize privacy as a moral or ethical value.

Integrity is a concept that Pirates understand, since it relates both to personal reputation in the community and freedom from suffocating legal norms. In addition, integrity informs gender relations among political geeks, making up part of the pirate vocabulary for moral-practical reasoning about acceptable conduct. "Geek feminism" adapts the concept for pirate politics. It wrestles with the glaring problems of sexism, heteronormativity, and underrepresentation of women in the intellectual and working lives of politically active pirates. Integrity thereby communicates a shared concern with autonomy while also appealing to a respect for the individuality and particularity of one's own private life.

Not surprisingly, women who are Pirate Party members or who engage in political discussion boards with Pirates have had to address a first-order problem among many of their male peers—namely, the stubborn insistence that sexism does not exist among politically unified Pirates. Among the more prominent Swedish feminists contributing ideas to pirate politics are Gameloft's Johanna Nylander (who wrote a blog called *File Sharing, Freedom, and Feminism*), Anna Troberg, and the journalist Isobel Hadley-Kamptz. Geek feminism injects desperately needed critiques of Pirate culture and broader political and social life into discourses about privacy and free speech. Geek sexism thrives in communities dominated by unseasoned and socially stunted young men, some of whom are unaware of their prejudices or are even proud to be sexist. Geek feminism generates ongoing, concrete examples of how women need stronger cyber-liberties protections, especially regarding privacy. The German Pirates, owing to their elected offices, have addressed women's issues somewhat more broadly than the Swedes. In 2012, the German Pirate Party endorsed an open letter criticizing Kristina Schroeder, the German minister for family and women, for ignoring structural factors contributing to sexism, such as Germany's high gender pay gap and inadequate day-care provisions (Dowling 2012).

Queer participation in pirate politics is well established, particularly in Germany, where LGBT rights were made explicit in the party platform in 2009 (Urbach, personal communication). Hacker equality generates ongoing discussions and debates about "the situation of queer people within the hacker and nerd community" (Socialhack 2012) at hacker conferences and conventions. Young queer nerds share concerns about bullying and finding community at panels at the annual Chaos Communication Conferences and Chaos Communication Camps and sponsor weekly "queer roundups" in Sweden and Germany. New music culture figures prominently in the formation of group ties among queer-trans pirates and hackers and attracts queer pirates internationally through Signal Internet radio shows at hackerspaces.org (Urbach, personal communication). Prominent hacker activists, including Meredeth L. Patterson and Rubin Starset, pursue projects promoting LGBT equality through hackerspace collectives like Noisebridge in San Francisco. As might be expected, queer and feminist hacker allegiances seem to elevate the technophilic aspect of personal identity above other aspects.

The visibly feminist and queer identity currents within pirate politics reinforce the hacker ethic of personal integrity and autonomy while also promoting antiauthoritarianism and freedom as social values. These hackers ultimately place universalism over particularistic ideologies and, especially, over Lundström-esque engagements with reactionary conservatism, racism, and nationalism.

The null hypothesis for pirate politics, as described previously, presses us to imagine that the emergence of cultural environmentalism is a temporary blip without lasting social or political consequence. Its conclusion is that pirate politics is nothing but a flash in the pan. But even if the Swedish and German Pirate

Parties were to disintegrate or disappear tomorrow (which is unlikely but not inconceivable), pirate politics would persist as a transnational cultural movement that urges the development of the Internet as a platform supporting democracy, shared knowledge, and governmental transparency. It would continue to contribute new Internet tools, vocabularies, and communities for pursuing those ends, independently of political parties and the identity politics of social movements. And it would continue to provide a social basis for challenging hermetic and illegitimate information policies with legal principles and procedures supportable by moral argumentation understandable by everybody. Quite possibly, these objectives can still become incorporated into the programs of more dominant political parties. In the meantime, the geeks and hackers happily rage on, semi-independently, emboldened by their newfound solidarity.

Notes

Introduction

1. To the best of my knowledge, "pirate politics" is a phrase coined by Jonas Andersson (see Andersson 2009b).

2. The "Celestial Jukebox" is the term Tom McCourt and I (Burkart and McCourt 2006), together with others including Vaidhyanathan (2005), use to discuss the plan for the Internet to become normalized into a mass medium for the production, distribution, and consumption of commercial media and other information commodities. The "Alternative Jukebox" (Burkart 2010) refers to the constellation of noncommercial media online and its cultural project.

3. The Greens/European Free Alliance's policy platform contains program initiatives that promote feminism, natural food, and linguistic diversity (G/EFA 2010).

4. A second Pirate member of the European Parliament from Sweden, twenty-two-year-old Amelia Andersdotter, was elected in 2009 after the Lisbon Treaty was signed by all member states of the EU (TorrentFreak 2009).

5. From the outside of the Berlin organization of the German Pirate Party, for example, it is difficult to identify either the presence of queer hacker culture among the activists, or the gender politics that accompanies many of the pirate and affiliated hacker conventions and plenaries.

6. BitTorrent is a decentralized file-sharing protocol that enables a file seeker to find and obtain entire files by compiling file parts from numerous sharing sources, provided that the seeker is also sharing files using the protocol; see BitTorrent 2012.

7. In 2009, following a Swedish court order, TPB closed for twenty-four hours, during which time users created more than 1,400 new BitTorrent sites to compensate (Espiner 2009).

8. Cultural reproduction, social integration, and socialization are life-world processes that correspond to the ongoing creation and maintenance of culture, society, and person, respectively; see Habermas 1981/2012.

9. For an example, see Escobar 2005. W. Lance Bennett (2012) updates many of these arguments for communication studies.

10. Author's note: "Geographical indications . . . are a type of intellectual property. They are forms of identification which identify a product as originating in a region or locality in a particular country" (EC 2011).

11. I distinguish another, related discourse by Touraine (1988) and his student Melucci (1989) that focused on "information age" characteristics of symbolic protests, rebellions over cultural "codes," and personal-identity politics. It complements resource-mobilization approaches (such as those of John D. McCarthy, Mayer N. Zald, and Charles Tilly) for studying protest movements by focusing on the symbolic politics of identity.

1   Nomads of the Information Society

1. The first members joined the party with a modest membership fee of 5 kronor paid via short message service, or SMS (*Local* 2006).

2. Falkvinge estimates that of the SPP members, "about two-thirds are creators of some sort, coders, musicians, writers, translators. The rest are mostly civil liberties activists" (qtd. in Raz 2009).

3. The Swedish Missionary Church of Kopimism, or Missionerande Kopimistsamfundet, promotes the cyberlibertarian beliefs associated

with free culture and claims to be inspired by a phrase in the Bible (Corinthians 11:1), modified slightly to "Copy me, my brothers, just as I copy Christ himself." Kopimism was formalized as an officially recognized religion in 2012 by the Kammarkollegiet agency.

4. Kelty (2008) and Coleman (2010) contribute important work on the lifeworld and cyberculture.

5. In Andersson's account, the pirates want to assert an alternative to the dualistic and juridical thinking about IP as free or as theft, and in the process, to shift some of the burdens of system integration left to the legal system today to the cultural commons.

6. See Kreisi 1989 for an overview of sociological studies of the new class (or, sometimes, "new middle class").

7. The Zapatistas ended their occupation in 2006, at the start of their transformation into a pure civil-society group ("la otra campaña," the other campaign). For more on Internet Zapatismo, see Rovira 2009, Russell 2005, and Ford and Gil 2001.

8. The fuzzy subjectivity of Anonymous is misrecognized by critics, many of whom disapprove of the use of distributed denial of service attacks that shut down communications, ostensibly in defense of free speech.

9. Vandystadt (2009a) frames the agenda as the intention to revise the information technology agreement in the World Trade Organization (WTO), to revise the EU telecoms package, and to influence the Stockholm Program on EU police cooperation. Evidence of ongoing agenda setting at http://breddning.piratpartiet.se/ reveals a broadening of the platform similar to the experience of the German Pirates. Martin Fredriksson discerns a persistent refusal to take a position on tax and benefits issues that could position them on a left-right continuum, and a willingness to commit to anti-nationalist positions on migration and to positions supporting gender and sexual equality (personal communication).

10. This pirate may be forgiven the rhetorical identification of freedom with koala bears, but "living space" is a much more unfortunate reference to Nazi expansionism.

11. Despite being a name that appears to be closely related to the Theory of Communicative Action, Bennett and Segerberg's description of a "logic of connective action" (2012) does not address pirate politics as an NSM. Their adoption of actor-network theory mistakes networking functionalities for social agency. Also, the underlying social and cultural conflicts underlying connective action are not explained with reference to colonization, but to anomie.

12. Schattschneider's (1988) account of the socialization of conflict has some purchase for pirate politics, just as it had for the U.S. civil rights era, for which it was originally developed.

13. German Pirates have criticized the Green Party for addressing gender underrepresentation in leadership through bureaucratic procedures that can jeopardize important votes. Specifically they have criticized the so-called 50–50 rule, which imposes a gender quota for Greens' rules of debate; this rule was blamed for lack of coordinated action on data retention and access-blocking laws in the Berlin state parliament.

14. The masks are the IP of Time Warner.

15. *Conviviality* refers to technology that fosters "the opposite of industrial productivity," namely, "autonomous and creative intercourse among persons . . . and with their environment," and "individual freedom realized in personal independence . . . [which is] an intrinsic ethical value" (Illich 1973, 11).

16. The Swedish technoculture and digital economy is partly shared with those of other Scandinavian countries. Nokia and Ericsson play influential roles, as do software makers using Linux.

17. As an EU member, Sweden is obliged to adopt the euro at some point (*Economist* 2003), but a 2003 referendum retained the krona as Sweden's national currency.

18. Finland had less need to publicly capitalize broadband development, relative to Sweden, in part because Finland's telecommunications development history has periods with more than one incumbent provider (Eskelinen, Frank, and Hirvonen 2008, 420).

19. Sweden has the world's second-largest proportionate share of ICT employment in the private sector, after Finland and ahead of Ireland (OECD 2009, 23).

20. For example, Peter Sunde, cofounder of TPB, created the micropayment company Flattr with Linus Olsson.

21. "The environmentalists helped us to see the world differently, to see that there was such a thing as 'the environment' rather than just my pond, your forest, his canal. We need to do the same thing in the information environment" (Boyle 2008, xv).

2 European Antipiracy Initiatives

1. As this book was going to press, ownership of Christiania was being transferred to a foundation run by its occupants for about $13 million (Thornburgh 2012).

Parts of this chapter were developed from Burkart and Logan 2013 and Burkart and Andersson 2012.

2. Charley McCreevy, the European Commissioner for Internal Market and Services, led an aggressive commission (the Internal Market and Services Directorate General, or DG MARKT) during the period of interest here. The mission of DG MARKT "is to secure . . . ever greater European market integration and to seek the removal of obstacles to the free movement of services and capital and to the freedom of establishment" of business activities (http://ec.europa.eu/commission_2010-2014/barnier/about/mandate/index_en.htm).

3. However, it was later rejected. For the text of IPRED, see http://ec .europa.eu/internal_market/iprenforcement/directive/index_en.htm. For the text of the Data Detention Directive, see http://eur-lex.europa .eu/LexUriServ/LexUriServ.do?uri=OJ:L:2006:105:0054:0063:EN:PDF.

4. While the commercial rationale for harmonization was evident at the time, an EC planning document ("Green Paper") titled "Combating Counterfeiting and Piracy in the Single Market" (1998) made the case for harmonization as a matter of public health and safety.

5. In this regard, two-level game theory using a rational-actor model of international politics predicts this outcome and explains it as a "win-win" for states with congruent interests (Putnam 1988). Game theory evades sociological concepts and is not addressed in this chapter.

6. The Framework Directive forms part of the "Telecommunications Package," together with the Authorization Directive, Access Directive, Universal Service Directive, and Privacy and Electronic Communications Directive.

7. The IMF considers cultural services to include audiovisual services delivered electronically: "Audiovisual and related services consist of services and fees related to the production of motion pictures (on film, videotape, disk, or transmitted electronically, etc.), radio and television programs (live or on tape), and musical recordings" (IMF 2009, 179–180).

8. In a report critical of the hybridized free-trade-IPR treaties, the Social Sciences Research Council claims that "the IIPA was instrumental in the creation of Special 301" review program (Karaganis 2010, 3).

9. Forum shifting prevents and forestalls sustained debate and discourse about the social desirability of the global IPR ratchet and permits non-linguistic steering media, principally lobbying money and asymmetrical international power relationships, to determine the outcomes of trade negotiations over IPR. Peter Drahos (2002) describes how, in multilateral negotiations, which are under the control of industry allies, the representatives of public interest and development communications are systematically ignored, sidelined, split apart, and overwhelmed with volumes of documentation generated by the negotiations. In negotiations, the logic of collective action for transnational industries that share an interest in protecting royalty-generating exports withstands the pressures of entropy and fragmentation better than the solidarity between the aligned developing countries and public interest groups.

10. It is noteworthy that the United States did not participate in the early phase of Berne Convention, "preferring to offer foreign authors little or no protection," since "American publishing was built on the piracy of European works" (Drahos 2002, 32–33). But the U.S. and Berne

Convention provisions were roughly similar except for Europe's recognition of moral rights. The United States ratified Berne in 1989 without recognizing moral rights.

11. For the text of the law, see, http://eur-lex.europa.eu/LexUriServ/LexUriServ.do?uri=CELEX:32001L0029:EN:HTML.

12. Brussels, 26.4.2006, COM(2006) 168 final. For the text of the law, see http://eur-lex.europa.eu/LexUriServ/LexUriServ.do?uri=COM:2006:0168:FIN:EN:HTML.

13. For the text of the law, see http://eur-lex.europa.eu/LexUriServ/LexUriServ.do?uri=CELEX:32006L0024:EN:HTML.

14. For the text of the law, see http://trade.ec.europa.eu/doclib/html/147937.htm.

15. It should be noted that the EU offers some important opportunities for community media and grassroots media to carve out noncommercial spaces for media producers (Jiménez and Scifo 2009).

16. For the text of the directive, see http://eur-lex.europa.eu/LexUriServ/LexUriServ.do?uri=CELEX:31996L0009:EN:HTML.

17. The text of the act is available at http://www.legislation.gov.uk/ukpga/2010/24/data.pdf.

3 Technoculture versus Big Brother

1. The right to permanently expunge personal profiles on social networks and other services is supported by Viviane Reding, EC commissioner for Information Society and Media, 2004–2010.

2. This formulation of the effect of colonization is taken from Eder 1995, 33.

3. Offe is critical of the change of emphasis in Marxist sociology from labor solidarity to NSMs; the identity-based movements emphasize difference and individualism at the price of universality and solidarity. His sociological insights and empirical analyses stand independently of these value judgments.

4. This prioritization still stands, despite the fact that the United Nations set a floor for digital rights in 2011 when it declared forcible disconnection to be a "violation of human rights" (Moya 2011).

5. I adopt Alvin Gouldner's concept of blocked ascendancy, which is also explored by Eder (2001).

6. Nonetheless, the SPP is demonized in mainstream political discourse. The IFPI and other industry groups portray the SPP as an illegal organization: "We are absolutely against the idea that any political party can give their support to the idea of free use of protected content" (qtd. in Masnick 2009a). U.S. representative Robert Wexler, former cochair of the Congressional Caucus on Intellectual Property Promotion and Piracy Prevention, lamented, "Government and private sector efforts to make IP theft taboo have fallen short" (qtd. in Quiggin 2009). Konstantin Kosachev, international affairs committee chairman of the Russian Duma, described a "dangerous crisis trend" in which "radical" political parties and "third wave forces" such as the Pirate Party would win seats in the European Parliament (TASS 2009).

7. The Green/European Free Alliance's policy platform contains program initiatives that promote feminist goals, natural food, and linguistic diversity (G/EFA 2010). The Pirates' alliance with the Green group was tested early (Taylor 2009).

8. Schramm's case began as a dispute about the terms of her commercial book deal, but has developed into a fuller exploration of the incompatibility of pirate politics with a "post-privacy" or "post-private" ideology. See Burkart and Andersson (forthcoming) on post-privacy as ideology.

References

AAP (Association of American Publishers). 2012. "BookStats Publishing Categories Highlights." http://publishers.org/bookstats/categories (accessed July 15, 2013).

Affigne, Anthony DeSales. 1990. International Impacts of Environmental Crisis. *National Forum* 70 (1) (Winter): 26–30.

Agarwal, Neeraj. 2010. Evaluating IPRED2: The Wrong Answer to Counterfeiting and Piracy. *Wisconsin International Law Journal* 27 (4): 790–816.

Agence France Presse. 2009. "Pirate Party Swashbuckles into Finnish Politics." August 19.

Andersson, Jonas. 2009a. "Q&A re 'Pirate Politics.'" *The Liquidculture Notebook.* http://liquidculture.wordpress.com/2009/10/13/qa-re-pirate-politics (accessed July 15, 2013).

Andersson, Jonas. 2009b. "Jonas Andersson and the Emergence of Pirate Politics." Interview by Michel Bauwens. P2P Foundation, October 31. http://blog.p2pfoundation.net/jonas-andersson-on-the-emergence-of-pirate-politics/2009/10/31 (accessed July 15, 2013).

Andersson, Jonas. 2011a. "The Origins and Impacts of Swedish Filesharing: A Case Study." *Critical Studies in Peer Production* 1. http://cspp.oekonux.org/research/mass-peer-activism (accessed July 15, 2013).

Andersson, Jonas. 2011b. "Peer-to-Peer-based File-Sharing beyond the Dichotomy of 'Downloading Is Theft' vs. 'Information Wants to Be Free': How Swedish File-Sharers Motivate Their Action." PhD diss., Goldsmiths, University of London.

Andrejevic, Mark. 2002. The Work of Being Watched: Interactive Media and the Exploitation of Self-Disclosure. *Critical Studies in Media Communication* 19 (2): 230–248.

Antipiratbyrån. 2012. "Om Antipiratbyrån." http://www.antipiratbyran .se (accessed July 15, 2013).

Axelsson, Ann-Sophie. 2010. Perpetual and Personal: Swedish Young Adults and Their Use of Mobile Phones. *New Media & Society* 12 (1): 35–54.

Barfield, Claude. 2011. "The Trans-Pacific Partnership: A Model for Twenty-First-Century Trade Agreements?" *American Enterprise Institute International Economic Outlook* 2. http://www.aei.org/files/2011/06/01/ IEO-2011-02-g.pdf (accessed July 15, 2013).

Baym, Nancy. 2011. The Swedish Model: Balancing Markets and Gifts in the Music Industry. *Popular Communication* 9 (1): 22–38.

BBC News. 2012. "ACTA: Germany Delays Signing Anti-piracy Agreement." February 10.

Beck, Ulrich. 1992. *The Risk Society: Towards a New Modernity*. London: Sage.

Becker, Sven. 2012. "Web Platform Makes Professor Most Powerful Pirate." *Der Spiegel*, February 3. http://www.spiegel.de/international/ germany/liquid-democracy-web-platform-makes-professor-most-power- ful-pirate-a-818683.html (accessed July 13, 2013).

Benkler, Yochai. 2006. *The Wealth of Networks: How Social Production Transforms Markets and Freedom*. New Haven, CT: Yale University Press.

Bennett, W. Lance. 2012. "The Logic of Connective Action: Digital Media and the Organization of Protest Politics." Virtual keynote address, International Communication Association, Phoenix, AZ, May 23–28.

Bennett, W. Lance, and Alexandra Segerberg. 2012. The Logic of Connective Action. *Information Communication and Society* 15 (5): 739–768.

BIS (Department for Business Innovation and Skills). 2010. *Digital Economy Act of 2010: Impact Assessments.*

BitTorrent. 2012. "Concepts." http://www.bittorrent.com/help/faq/concepts (accessed July 15, 2013).

Boyle, James. 2008. *The Public Domain: Enclosing the Commons of the Mind.* New Haven, CT: Yale University Press.

Braman, Sandra. 2004. The Emergent Global Information Policy Regime. In *The Emergent Global Information Policy Regime*, ed. Sandra Braman, 12–39. New York: Palgrave.

Braman, Sandra. 2006. *Change of State: Information, Policy, and Power.* Cambridge, MA: MIT Press.

Buechler, Steven M. 1995. New Social Movement Theories. *Sociological Quarterly* 36 (3): 441–464.

Burkart, Patrick. 2010. *Music and Cyberliberties.* Middletown, CT: Wesleyan University Press.

Burkart, Patrick, and Jonas Andersson. 2012. "Inside the ACTA Negotiations: A Stakeholder Enumeration." Paper presented at "Communication and Media Policy in the Era of the Internet and Digitization," a workshop of the Communication Law and Policy Section of the European Communication Research and Education Association. Munich, March 16.

Burkart, Patrick, and Jonas Andersson. Forthcoming. Post-privacy and Ideology. In *Media, Surveillance and Identity.* Pieterlen, Switzerland: Peter Lang AG.

Burkart, Patrick, and L. Logan. 2013. Media Production and Information Policy: Growth through Replication. In *International Companion to Media Studies*, vol. 2, ed. Angharad Valdivia, 61–82. Hoboken, NJ: Blackwell.

Burkart, Patrick, and Tom McCourt. 2006. *Digital Music Wars: Ownership and Control of the Celestial Jukebox.* Lanham, MD: Rowman and Littlefield.

Cadwalladr, Carole. 2012. "Rick Falkvinge: The Swedish Radical Leading the Fight over Web Freedoms." *Observer*, January 22. http://www.guardian.co.uk/technology/2012/jan/22/rick-falkvinge-swedish-radical-web-freedoms (accessed July 15, 2013).

Calabrese, Andrew, and Marco Briziarelli. 2011. Policy Imperialism: Bilateral Trade Agreements as Instruments of Media Governance. In *The Handbook of Global Media and Communication Policy*, ed. Robin Mansell and Marc Raboy, 383–394. Malden, MA: Blackwell.

Caldwell, Christopher. 2009. Steal This eBook: An Internet Piracy Party Grows in Sweden. *Weekly Standard*, June 29.

Calhoun, Craig. 1993. "New Social Movements" of the Early Nineteenth Century. *Social Science History* 17 (3): 385–427.

Calhoun, Craig. 1995. *Critical Social Theory: Culture, History, and the Challenge of Difference*. Cambridge, MA: Wiley-Blackwell.

Carlsson, Bo. 1995. Communicative Rationality and Open-Ended Law in Sweden. *Journal of Law and Society* 22 (4): 475–505.

Carr, Daphne. 2011. "Czech Free Culture/Svoboda Cultural." Paper presented at the International Association for the Study of Popular Music, Cincinnati, OH.

Carter, Richard. 2009. "German Officials Probe Twitter Election 'Leaks.'" Agence France-Presse, September 28.

Castells, Manuel. 2001. *The Internet Galaxy*. New York: Oxford University Press.

Castells, Manuel. 2010. *The Rise of the Network Society*, vol. 1 of *The Information Age: Economy, Society, and Culture*, 2nd ed. Malden, MA: Wiley-Blackwell.

Chu, Henry. 2009. "A Declaration of Web Independence; Swedish Youths Back a Budding Political Party Based on a Single Issue." *Los Angeles Times*, December 27.

Clarke, Jim. 2009. "Pirate Party to Invade Ireland." *Sunday Mirror*, June 21.

Cohen, Jean L. 1982. *Class and Civil Society: The Limits of Marxian Critical Theory*. Amherst: University of Massachusetts Press.

Cohen, Jean L. 1985. Strategy or Identity: New Theoretical Paradigms and Contemporary Social Movements. *Social Research* 52 (4): 663–716.

Cohen, Jean L., and Andrew Arato. 1992. *Civil Society and Political Theory*. Cambridge, MA: MIT Press.

Cohen, Mitchell. 1992. Rooted Cosmopolitanism: Thoughts on the Left, Nationalism, and Multiculturalism. *Dissent* 39 (4): 478–483.

Colby, Kevin T. 1988. Public Access to Land: Allemansrätt in Sweden. *Landscape and Urban Planning* 15:253–264.

Coleman, Gabriella. 2004. The Political Agnosticism of Free and Open Source Software and the Inadvertent Politics of Contrast. *Anthropological Quarterly* 77 (3): 507–519.

Coleman, Gabriella. 2010. The Hacker Conference: A Ritual Condensation and Celebration of a Lifeworld. *Anthropological Quarterly* 83 (1): 47–72.

Computer Weekly. 2009. "Unmask Swetorrents File Sharers or Face Fine, ISP Told." *Computer Weekly*, December 7. http://www.computerweekly.com/news/1280091583/Unmask-Swetorrents-file-sharers-or-face-fine-ISP-told (accessed July 15, 2013).

Copycense. 2009. "Foreign Affairs as the New Copyright Law." *Copycense*, June 2. Part 2 of 3. http:// http://copycense.com/2009/06/09/foreign_affairs_piracy_legislation/ (accessed July 13, 2013).

Crossley, Nick. 2003. Even Newer Social Movements? Anti-Corporate Protests, Capitalist Crises and the Remoralization of Society. *Organization* 10 (2): 287–305.

CTK (CTK National News Wire). 2009. "Swedish Piratpartiet-like Party Emerging in Slovakia." June 11.

Curran, Giorel, and Morgan Gibson. 2012. WikiLeaks, Anarchism, and Technologies of Dissent. *Antipode*. doi:10.1111/j.1467-8330.2012.01009.x (accessed July 15, 2013).

Czech News Agency. 2010. "Czech Mainstream Parties Unattractive for Youth–Experts." May 14.

Dalton, Russell J., Manfred Kuechler, and Wilhelm Bürkli. 1990. The Challenge of New Movements. In *Challenging the Political Order: New Social and Political Movements in Western Democracies*, ed. Russell J. Dalton and Manfred Kuechler, 3–20. New York: Oxford University Press.

Della Porta, Donatella, and Sidney Tarrow, eds. 2005. *Transnational Protest and Global Activism*. Lanham, MD: Rowman and Littlefield.

Demers, Joanna. 2006. *Steal This Music: How Intellectual Property Law Affects Musical Creativity*. Athens: University of Georgia Press.

Deutsche Presse-Agentur. 2009. "Slovaks, Czechs Forming Their Own Pirate Parties." June 11.

Deutsche Welle. 2012. "Sweden Passes Controversial Data Retention Directive." March 22. http://www.dw.de/sweden-passes-controversial -data-retention-directive/a-15826462 (accessed July 13, 2013).

Dews, Peter. 1986. *Autonomy and Solidarity: Interviews with Jürgen Habermas*. London: Verso.

Dietz, Jan L. G. 1991. Speech Acts or Communicative Action? In *Proceedings of the Second European Conference on Computer-Supported Cooperative Work*, ed. Liam Bannon, Mike Robinson, and Kjeld Schmidt, 235–248. Dordrecht: Kluwer.

Doctorow, Cory. 2005. "Why Some 'Piracy' Can Increase Overall Revenues." BoingBoing, August 24. http://www.boingboing.net/2005/08/24/ why-some-piracy-can-html (July 13, 2013).

Domscheit-Berg, Daniel. 2010. Interview in *WikiRebels: The Documentary*, directed by Jesper Huor and Bosse Lindquist for Sveriges Television AB.

Dowling, Siobhan. 2012. "Women Are Doing Fine, Says Germany's Minister for Women." GlobalPost, May 6. http://www.globalpost.com/ dispatch/news/regions/europe/germany/120503/germany-minister -women-kristina-schroeder-gender-debate (accessed July 15, 2013).

Downing, John. 1996. *Internationalizing Media Theory: Transition, Power, Culture.* Thousand Oaks, CA: Sage.

Downing, John. 2001. *Radical Media: Rebellious Communication and Social Movements.* Thousand Oaks, CA: Sage.

Drahos, Peter. 2002. *Information Feudalism: Who Owns the Knowledge Economy?* London: Earthscan.

Dreyfuss, Rochelle Cooper. 2004. TRIPS-Round II: Should Users Strike Back? *University of Chicago Law Review.* 71 (1): 21–35.

EC (European Commission). 2005a. "Media Pluralism: What Should Be the European Union's Role?" Information Society and Media Directorate-General. Issues paper for the Liverpool Audiovisual Conference. July. http://ec.europa.eu/avpolicy/docs/reg/modernisation/issue_papers/ ispa_mediaplur_en.pdf (accessed July 15, 2013).

EC (European Commission). 2005b. "Electronic Communications: The Road to the Knowledge Economy." October 25. http://europa.eu/ legislation_summaries/other/l24216b_en.htm (accessed July 13, 2013).

EC (European Commission). 2011. "Geographical Indications." September 1. http://ec.europa.eu/trade/policy/accessing-markets/intellectual-property/geographical-indications/ (accessed July 15, 2013).

ECDGT (European Commission Directorate-General for Trade). 2010. "Intellectual Property: Anti-Counterfeiting." http://trade.ec.europa.eu/ doclib/press/index.cfm?id=552 (accessed July 15, 2013).

EC-MARKT (European Commission, Internal Market and Services). 2010. "Draft Summary: Stakeholders' Dialogue on Illegal Up- and Downloading, 1 July 2010." http://www.scribd.com/doc/38532573/ Annex-6-Draft-Summary-Record-Stakeholders-Dialogue-Piracy -01072010-PCInpact (accessed July 15, 2013).

Economist. 2003. "Keeping the Krona." September 15.

Economist. 2004. "Get Orf My Land!" September 18, 62.

Edelman, Murray Jacob. 1971. *Politics as Symbolic Action: Mass Arousal and Quiescence.* Chicago: Markham.

Eder, Klaus. 1985. The "New Social Movements": Moral Crusades, Political Pressure Groups, or Social Movements? *Social Research* 52 (4) (Winter): 869–890.

Eder, Klaus. 1988. *The Social Construction of Nature: A Sociology of Ecological Enlightenment*. Trans. Mark Ritter. Thousand Oaks, CA: Sage.

Eder, Klaus. 1995. Does Social Class Matter in the Study of Social Movements? A Theory of Middle-Class Radicalism. In *Social Movements and Social Classes: The Future of Collective Action*, ed. Louis Maheu, 21–53. Thousand Oaks, CA: Sage.

Eder, Klaus. 1996. *New Politics of Class: Social Movements and Cultural Dynamics in Advanced Societies*. Thousand Oaks, CA: Sage.

Eder, Klaus. 2001. Sustainability as a Discursive Device for Mobilizing European Publics. In *Environmental Politics in Southern Europe: Actors, Institutions, and Discourses in a Europeanizing Society*, ed. Klaus Eder and Maria Kousis, 25–52. Norwell, MA: Kluwer.

Eder, Klaus. 2007. The Public Sphere and European Democracy: Mechanisms of Democratization in the Transnational Situation. In *The European Union and the Public Sphere: A Communicative Space in the Making?*, John Erik Fossum and Philip Schlesinger, 44–64. New York: Routledge.

Eder, Klaus, and Maria Kousis, eds. 2001. *Environmental Politics in Southern Europe: Actors, Institutions, and Discourses in a Europeanizing Society*. Norwell, MA: Kluwer.

EDRI (European Digital Rights Initiative). 2011. "German Internet Blocking Law to Be Withdrawn." April 6. http://www.edri.org/edrigram/number9.7/germany-internet-blocking-law (accessed July 15, 2013).

Edwards, Chris. 2009. "Technology: Sweden's Pirate Party Sails to Success in European Elections." *Guardian*, June 11.

Edwards, Gemma. 2004. Habermas and Social Movements: What's "New?" *Sociological Review* 52 (1): 113–130.

Edwards, Gemma. 2008. The "Lifeworld" as a Resource for Social Movement Participation and the Consequences of Its Colonization. *Sociology* 42 (2): 299–316.

EFF (Electronic Frontier Foundation). 2007. "IPRED2 after the Committee." March 21. https://www.eff.org/deeplinks/2007/03/ipred2-after-committee (accessed July 15, 2013).

EIU (Economist Intelligence Unit). 2011. "Sweden Economy: North Star." ViewsWire, June 11.

Elkin-Koren, Niva. 2000. The Privatization of Information Policy. *Ethics and Information Technology* 2:201–209.

Engström, Christian. 2009a. "Copyright Laws Threaten Our Online Freedom." *Financial Times*, July 7.

Engström, Christian. 2009b. "Distribution Capacity Is in Everyone's Hands." Interview by Nathalie Vandystadt. *Europolitics*, September 7. http://www.europolitics.info/distribution-capacity-is-in-everyone-s-hands-art246888-1.html (accessed July 15, 2013).

Engström, Christian. 2012a. "Pirate Party Platform for the 2009 EU Elections." January 5. http://christianengstrom.wordpress.com/2012/01/05/pirate-party-platform-for-the-2009-eu-elections/ (accessed July 15, 2013).

Engström, Christian. 2012b. "ACTA: Victory!" July 4. https://christianengstrom.wordpress.com/2012/07/04/acta-victory/ (accessed July 15, 2013).

EP (European Parliament). 2000. "The European Parliament and the World Trade Organization (WTO/GATT)." European Parliament Fact Sheets, June 29. http://www.europarl.europa.eu/factsheets/6_2_2_en.htm (accessed July 15, 2013).

EP (European Parliament). 2009. "Access to Documents: The European Parliament Demands More Transparency." March 11. http://www.europarl.europa.eu/sides/getDoc.do?pubRef=-//EP//TEXT+IM-PRESS+20090310IPR51408+0+DOC+XML+V0//EN (accessed July 15, 2013).

ESA (Entertainment Software Association). 2012. "Industry Facts." http://www.theesa.com/facts/index.asp (accessed July 15, 2013).

Escobar, Arturo. 2005. "Other Worlds Are (Already) Possible: Cyber-Internationalism and Post-Capitalist Cultures." *Textos de la CiberSociedad* 5. http://www.cibersociedad.net/textos/articulo.php?art=18 (accessed July 15, 2013).

Eskelinen, Heikki, Lauri Frank, and Timo Hirvonen. 2008. Does Strategy Matter? A Comparison of Broadband Rollout Policies in Finland and Sweden. *Telecommunications Policy* 32:412–421.

Espiner, Tom. 2009. "Pirate Bay Closure Sparks Rise in P2P Sites." ZDNet, November 3. http://www.zdnet.com/pirate-bay-closure-sparks-rise-in-p2p-sites-3039854663 (accessed July 15, 2013).

EurActiv. 2010. "Parliament Threatens Court Action on Anti-piracy Treaty." March 12. http://www.euractiv.com/health/meps-defy-commission-internet-piracy-agreement-news-326215 (accessed July 15, 2013).

EuroNews. 2009. "Young German Voters Turn to Fringe Parties." September 21. http://www.euronews.com/2009/09/21/young-german-voters-turn-to-fringe-parties/ (accessed July 15, 2013).

Expo Idag. 2005. "Carl Lundström, Mångmiljonär och Högerextremist." March 3. http://expo.se/2005/carl-lundstrom,-mangmiljonar-och-hogerextremist_1289.html (accessed July 15, 2013).

Eyerman, Ron, and Andrew Jamison. 1998. *Music and Social Movements: Mobilizing Traditions in the Twentieth Century.* New York: Cambridge University Press.

Falkvinge, Rickard. 2011. "Cable Reveals Extent of Lapdoggery from Swedish Govt on Copyright Monopoly." September 5. http://falkvinge.net/2011/09/05/cable-reveals-extent-of-lapdoggery-from-swedish-govt-on-copyright-monopoly/ (accessed July 15, 2013).

Fehér, Ferenc, and Agnes Heller. 1983. From Red to Green. *Telos* 59:35–44.

FFII. (Foundation for a Free Information Infrastructure). 2005. "Rozmanith: Using Software Patents to Silence Critics." http:// http://eupat.ffii.org/pikta/xrani/rozmanith/index.en.html (accessed July 13, 2013).

FFII. 2010. "ACTA Will Undermine European Parliament's Power in IP Matters." http://press.ffii.org/Press%20releases/ACTA%20will %20undermine%20European%20Parliament's%20power%20in%20IP %20matters (accessed July 15, 2013).

FIPR (Foundation for Information Policy Research). 2005. "The Second IPR Enforcement Directive: A Threat to Competition and Liberty." http://www.fipr.org/copyright/ipred2.html (accessed July 15, 2013).

Fiveash, Kelly. 2009. "Swedish Web Traffic Tumbles as IPRED Arrives." *Register,* April 2. http://www.theregister.co.uk/2009/04/02/sweden _internet_traffic_falls_ipred (accessed July 15, 2013).

Forbes. 2012. "Ericsson." Global 2000 Leading Companies. http://www .forbes.com/companies/ericsson (accessed July 15, 2013).

Ford, Tamara, and Genève Gil. 2001. Radical Internet Use. In *Radical Media: Rebellious Communication and Social Movements*, ed. John Downing, 201–234. Thousand Oaks, CA: Sage.

Fossum, John Erik, and Philip Schlesinger, eds. 2007. *The European Union and the Public Sphere: A Communicative Space in the Making?* New York: Routledge.

Fraser, Nancy. 1985. What's Critical about Critical Theory? The Case of Habermas and Gender. *New German Critique,* 35:97–132.

GamePolitics. 2010. "USTR Issues Special 301 Report on Global IP Enforcement." May 3. http://www.gamepolitics.com/2010/05/03/ustr-issues -special-301-report-global-ip-enforcement (accessed July 15, 2013).

Gasser, Urs, and Michael Girsberger. 2004. "Transposing the Copyright Directive: Legal Protection of Technological Measures in EU-Member States." Berkman Publication Series no. 2004–10. Berkman Center for Internet and Society, Harvard Law School. http://cyber.law.harvard.edu/ media/files/eucd.pdf (accessed July 15, 2013).

GCN (Government Computer News). 2012. "EU Signs 'International SOPA' as Online Protests Continue." January 26. http://gcn .com/articles/2012/01/26/anonymous-leads-ddos-attacks-in-europe .aspx (accessed July 15, 2013).

G/EFA (Greens/European Free Alliance). 2010. The Greens/European Free Alliance in the European Parliament. http://www.greens-efa.eu (accessed July 15, 2013).

Geist, Michael. 2003. Cyberlaw 2.0. *Boston College Law Review* 44 (2): 323–358.

Geist, Michael. 2010. "ACTA Ultra-Lite: The U.S. Cave on the Internet Chapter Complete." *ACTA Watch*, October 6. http://acta.michaelgeist .ca/blog/acta-ultra-lite-us-cave-internet-chapter-complete (accessed July 15, 2013).

Gibler, John. 2008. *Mexico Unconquered: Chronicles of Power and Revolt.* San Francisco: City Lights.

Gillespie, Tarleton. 2009. Characterizing Copyright in the Classroom: The Cultural Work of Antipiracy Campaigns. *Communication, Culture & Critique* 2 (3): 274–318.

GlobalComms (GlobalComms Database Service). 2012. "Sweden." Subscription service (accessed June 5, 2012).

Gollmitzer, Mirjam. 2008. Industry versus Democracy: The New "Audiovisual Media Services Directive" as a Site of Ideological Struggle. *International Journal of Media and Cultural Politics* 4 (3): 331–348.

Gouldner, Alvin Ward. 1979. *The Future of Intellectuals and the Rise of the New Class.* New York: Seabury.

Gouldner, Alvin Ward. 1985. *Against Fragmentation: The Origins of Marxism and the Sociology of Intellectuals.* New York: Oxford University Press.

Greenberg, Andy. 2011. "Iceland's Data Haven Plan Finds an American Friend: Internet Archive's Brewster Kahle." *Forbes*, February 1. http://www.forbes.com/sites/andygreenberg/2011/02/01/icelands-data -haven-plan-finds-an-american-friend-internet-archives-brewster-kahle (accessed July 15, 2013).

Habermas, Jürgen. 1981. New Social Movements. *Telos* 49:33–37.

Habermas, Jürgen. 1984. *Reason and the Rationalization of Society*, vol. 1 of *The Theory of Communicative Action*, trans. Thomas McCarthy. Boston: Beacon.

Habermas, Jürgen. 1987a. *Lifeworld and System: A Critique of Functionalist Reason*, vol. 2 of *The Theory of Communicative Action*, trans. Thomas McCarthy. Boston: Beacon.

Habermas, Jürgen. 1987b. *The Philosophical Discourse of Modernity: Twelve Lectures*. Cambridge, MA: MIT Press.

Habermas, Jürgen. 1988. Legitimation Crisis. Trans. Thomas McCarthy. Cambridge, MA: Polity Press. First published by Beacon Press, Boston, 1975.

Habermas, Jürgen. 1996. *Between Facts and Norms*. Cambridge, MA: MIT Press.

Habermas, Jürgen. 2001. *The Postnational Constellation: Political Essays*. Trans. Max Pensky. Cambridge, MA: Polity Press.

Habermas, Jürgen. 2010. "Leadership and Leitkulture." *New York Times*, October 28.

Habermas, Jürgen. 1981/2012. Modernity: An Unfinished Project. In *Contemporary Sociological Theory*, ed. Craig Calhoun, Joseph Gerteis, James Moody, Steven Pfaff, and Indermohan Virk, 444–450. Malden, MA: Wiley-Blackwell.

Hartley, Matt. 2009. "Internet Freedom Movement Wins EU Seat: New Party Advocates Stronger Online Privacy." *Toronto Globe and Mail*, June 9.

Havel, Václav. 1991. Politics and Conscience (1984). In *Open Letters*, 249–271. London: Faber and Faber.

Heller, Christian. 2011. *Post-Privacy: Prima Leben Ohne Privatsphäre*. Munich: Beck.

Henderson, Karen. 2009. Europeanization of Political Parties: Redefining Concepts in a United Europe. *Sociologica* 41 (6): 526–538.

Herman, Bill D., and Oscar H. Gandy. 2006. Catch 1201: A Legislative History and Content Analysis of the DMCA Exemption Proceedings. *Cardozo Arts and Entertainment Law Journal* 24:121–190.

Hetherington, Kevin. 1998. *Expressions of Identity: Space, Performance, Politics*. Thousand Oaks, CA: Sage.

Hilgartner, Stephen. 2009. Intellectual Property and the Politics of Emerging Technology: Inventors, Citizens, and Powers to Shape the Future. *Chicago-Kent Law Review* 84 (1): 197–222.

Hinze, Gwen. 2004. "Proposed E.U. Directive on Intellectual Property Enforcement." Electronic Freedom Foundation brief. http://www.eff.org/files/filenode/effeurope/EU_IPRED_analysis.pdf (accessed July 13, 2013).

Hix, Simon, Abdul Noury, and Gérard Roland. 2006. Dimensions of Politics in the European Parliament. *American Journal of Political Science* 50 (2): 494–511.

Honneth, Axel. 1979. Communication and Reconciliation: Habermas' Critique of Adorno. *Telos* 39:45–61.

Humeau, Marie. 2011. "Sweden Argues That Transposing Data Retention Directive Is Unnecessary." European Digital Rights, September 7. http://www.edri.org/edrigram/number9.17/sweden-contests-data-retention-unnecessary (accessed July 15, 2013).

Hyde, Lewis. 2010. *Common as Air: Revolution, Art, and Ownership*. New York: Farrar, Straus and Giroux.

IFTA (Independent Film and Television Alliance). 2012. "About IFTA." http://www.ifta-online.org/about-ifta (accessed July 15, 2013).

Iida, Keisuke. 2004. Is WTO Dispute Settlement Effective? *Global Governance* 10:207–225.

IIPA (International Intellectual Property Alliance). 2007. "Member Associations." http://www.iipa.com/memberassociations.html (accessed July 15, 2013).

IIPA (International Intellectual Property Alliance). n.d. "Description of the IIPA." http://www.iipa.com/aboutiipa.html (accessed July 15, 2013).

Illich, Ivan. 1973. *Tools for Conviviality*. New York: Harper and Row.

IMF (International Monetary Fund). 2009. *Balance of Payments and International Investment Position Manual.* 6th ed. Washington, DC: International Monetary Fund.

Ingdahl, Waldemar. 2010. Case Study: The Pirate Code on Trial in Sweden—What Future for Intellectual Property? In *International Property Rights Index: 2010 Report*, ed.Victoria Strokova, 64–66. Potsdam-Babelsberg: Liberal Institute of the Friedrich Naumann Foundation; Berlin: Institute for Free Enterprise.

Inskeep, Steve. 2009. "Pirate Party Does Well in European Election." National Public Radio, *Morning Edition*, June 12. http://www.npr.org/templates/story/story.php?storyId=105285841 (accessed July 15, 2013).

Jeffrey, Craig. 2012. "Geographies of Children and Youth III: Alchemists of the Revolution?" *Progress in Human Geography*, January 25. Online publication ahead of print. doi:10.1177/0309132511434902 (accessed July 15, 2013).

Jiménez, Núria Reguero, and Salvatore Scifo. 2009. Community Media in the Context of European Media Policies. *Telematics and Informatics* 27 (2): 131–140.

Johanssen, Dan. 2004. Is Small Beautiful? The Case of the Swedish IT Industry. *Entrepreneurship and Regional Development* 16 (July): 271–287.

Johns, Adrian. 2010. *Piracy: The Intellectual Property Wars from Gutenberg to Gates.* Chicago: University of Chicago Press.

Jones, Ben. 2011. "Lib Dems to Vote on Piracy Act Repeal." TorrentFreak, August 26. http://torrentfreak.com/lib-dems-to-vote-on-piracy-act-repeal-110826/ (accessed July 15, 2013).

Jones, K. C. 2007. "Anti-copyright Pirate Party Seeks Official Recognition." *InformationWeek*, August 13. http://www.informationweek.com/anti-copyright-pirate-party-seeks-offici/201500109 (accessed July 13, 2013).

Jordan, Tim, and Paul Taylor. 2004. *Hacktivism and Cyberwars: Rebels with a Cause.* New York: Routledge.

Karaganis, Joe. 2010. Letter to Jennifer Choe Groves, Office of the United States Trade Representative, re: 2010 Special 301 Review, Docket Number USTR-2010–0003. Social Science Research Council, February 18. http://piracy.americanassembly.org/wp-content/uploads/2010/12/2010-Special-301-SSRC-Comment.pdf (accessed July 13, 2013).

Keck, Margaret E., and Kathryn Sikkink. 1998. *Activists without Borders: Advocacy Networks in International Politics*. Ithaca, NY: Cornell University Press.

KEI (Knowledge Ecology International). 2012. "European Parliament Votes 478 to 39 to Reject ACTA." http://keionline.org/node/1454 (accessed July 15, 2013).

Kellner, Douglas. 1989. *Critical Theory, Marxism, and Modernity*. Cambridge: Polity Press.

Kelly, Christine A. 2001. *Tangled up in Red, White, and Blue: New Social Movements in America*. Lanham, MD: Rowman and Littlefield.

Kelty, Christopher. 2005. Geeks, Social Imaginaries, and Recursive Publics. *Cultural Anthropology* 20 (2): 185–214.

Kelty, Christopher. 2008. *Two Bits: The Cultural Significance of Free Software*. Durham, NC: Duke University Press.

Kennedy, Michael D. 2011. "Global Solidarity and the Occupy Movement." *Possible Futures*, December 5. http://www.possible-futures.org/2011/12/05/global-solidarity-occupy-movement (accessed July 15, 2013).

Kousis, Maria, and Klaus Eder. 2001. EU Policy-Making, Local Action, and the Emergence of Institutions of Collective Action. In *Environmental Politics in Southern Europe: Actors, Institutions, and Discourses in a Europeanizing Society*, ed. Klaus Eder and Maria Kousis, 3–21. Norwell, MA: Kluwer.

Kravets, David. 2010. "ACTA Backs Away from 3 Strikes." *Wired*, April 21.

Kreisi, Hanspeter. 1989. New Social Movements and the New Class in the Netherlands. *American Journal of Sociology* 94 (5): 1078–1116.

Larsson, Anders Olof. 2011. "'Extended Infomercials' or 'Politics 2.0'? A Study of Swedish Political Party Web Sites before, during and after the 2010 Election." *First Monday* 16, no. 4. http://firstmonday.org/ojs/index .php/fm/article/view/3456/2858 (accessed July 13, 2013).

Larsson, Stefan. 2011. The Path Dependence of European Copyright. *scripted* 8 (1): 8–31.

Larsson, Stefan, and Måns Svensson. 2010. Compliance or Obscurity? Online Anonymity as a Consequence of Fighting Unauthorised File-sharing. *Policy and Internet* 2 (4). doi:10.2202/1944-2866.1044 (accessed July 15, 2013).

LaRue, Frank. 2011. "Report of the Special Rapporteur on the Promotion and Protection of the Right to Freedom of Opinion and Expression." UN General Assembly, Human Rights Council, A/HRC/17/27. May 16. http://www2.ohchr.org/english/bodies/hrcouncil/docs/17session/A .HRC.17.27_en.pdf (accessed July 15, 2013).

Layton, Edwin T. 1986. *The Revolt of the Engineers: Social Responsibility and the American Engineering Profession.* Baltimore, MD: Johns Hopkins University Press.

Lessig, Lawrence. 1999. *Code and Other Laws of Cyberspace.* Boston: Basic Books.

Lessig, Lawrence. 2002. *The Future of Ideas: The Fate of the Commons in a Connected World.* New York: Vintage.

Li, Miaoran. 2009. The Pirate Party and The Pirate Bay: How The Pirate Bay Influences Sweden and International Copyright Relations. *Pace International Law Review* 21 (1): 281–307.

Libbenga, Jan. 2007. "The Pirate Bay Plans to Buy Sealand." *Register,* January 12. http://www.theregister.co.uk/2007/01/12/pirate_bay_buys _island (accessed July 15, 2013).

Lim, Merlyna, and Mark E. Kann. 2008. Politics: Deliberation, Mobilization, and Networked Practices of Agitation. In *Networked Publics,* ed. Kazys Varnelis, 77–108. Cambridge, MA: MIT Press.

Lischka, Von Konrad. 2007. "Rechtspopulist finanziert Internet-Taus-chbörse." *Der Spiegel*, April 5. http://www.spiegel.de/netzwelt/web/piratenseite-im-zwielicht-rechtspopulist-finanziert-internet-tauschboerse-a-480972.html (accessed July 15, 2013).

Löblich, Maria, and Manuel Wendelin. 2012. "ICT Policy Activism on a National Level: Ideas, Resources, and Strategies of German Civil Society in Governance Processes." *New Media and Society*. Online publication ahead of print. doi:10.1177/1461444811432427 (accessed July 15, 2013).

Local. 2006. "Pirate Party 'Larger than Greens.'" August 8. http://www.thelocal.se/4542/20060808 (accessed July 15, 2013).

Local. 2008. "Swedish Surveillance Law 'Breaks EU Rules.'" August 13. http://www.thelocal.se/13664/20080813 (accessed July 15, 2013).

Love, James. 2009. "White House Shares the ACTA Internet Text with 42 Washington Insiders, under Non Disclosure Agreements." Knowledge Ecology International, October 13. http://keionline.org/node/660 (accessed July 15, 2013).

LQN (La Quadruture du Net). 2011. "WikiLeaks Cables Shine Light on ACTA History." February 3. http://www.laquadrature.net/en/wikileaks-cables-shine-light-on-acta-history (accessed July 15, 2013).

LQN (La Quadruture du Net). 2012. "ACTA Rapporteur Denounces ACTA Masquerade." January 28. https://www.laquadrature.net/wiki/ACTA_rapporteur_denounces_ACTA_mascarade (accessed July 15, 2013).

Marketline. 2011. "Global Software." December. http://www.reportlinker.com/p0188773-summary/Global-Software.html (accessed July 15, 2013).

Marsden, Christopher T. 2011. *Internet Co-Regulation: European Law, Regulatory Governance, and Legitimacy in Cyberspace*. Cambridge: Cambridge University Press.

Masnick, Mike. 2009a. "Once Again Privacy Laws and Anti-piracy Data Retention Laws Conflict." Techdirt, May 26. http://www.techdirt.com/articles/20090521/1821104967.shtml (accessed July 15, 2013).

Masnick, Mike. 2009b. "Compare the Process between Engstrom's Internet Bill of Rights and ACTA." Techdirt, December 10. http://www.techdirt.com/articles/20091209/1850187280.shtml (accessed July 15, 2013).

Masnick, Mike. 2010a. "EU Has a 'Public/Private' IP Observatory to Watch for Copyright Infringement Online." Techdirt, February 12. http://www.techdirt.com/articles/20100209/0441358095.shtml (accessed July 15, 2013).

Masnick, Mike. 2010b. "Swedish Officials Complained to US That Hollywood-Pushed IPRED 'Anti-Piracy' Law Did More Harm than Good." Techdirt, December 27. http://www.techdirt.com/articles/20101226/00231112409/swedish-officials-complained-to-us-that-hollywood-pushed-ipred-anti-piracy-law-did-more-harm-than-good.shtml (accessed July 15, 2013).

McCullough, Malcolm. 2006. On Urban Markup: Frames of Reference in Location Models for Participatory Urbanism. *Leonardo Electronic Almanac* 14 (3) (July), 1–5.

McDonald, Kevin. 2006. *Global Movements: Action and Culture*. Malden, MA: Blackwell.

McLeod, Kembrew. 2001. *Owning Culture: Authorship, Ownership, and Intellectual Property Law*. New York: Lang.

McLeod, Kembrew. 2005. *Freedom of Expression®: Overzealous Copyright Bozos and Other Enemies of Creativity*. New York: Doubleday.

Mead, George Herbert. 1962. *Mind, Self, and Society from the Standpoint of a Social Behaviorist*, vol. 1. Chicago: University of Chicago Press.

Meiksins, Peter. 1988. The "Revolt of the Engineers" Reconsidered. *Technology and Culture* 29 (2): 219–246.

Meller, Paul. 2009. "Sweden Aims High for Creation of a Single EU Patent System." PCWorld, July 2. http:// http://www.pcworld.com/article/167772/article.html (accessed July 13, 2013).

Melucci, Alberto. 1989. *Nomads of the Present: Social Movements and Individual Needs in Contemporary Society*. Philadelphia: Temple University Press.

Melucci, Alberto. 1995. The New Social Movements Revisited: Reflections on a Sociological Misunderstanding. In *Social Movements and Social Classes: The Future of Collective Action*, ed. Louis Maheu, 107–119. Thousand Oaks, CA: Sage.

Meunier, Sophie, and Kalypso Nicolaïdis. 2006. The European Union as a Conflicted Trade Power. *Journal of European Public Policy* 13 (6): 906–925.

Meza, Ed. 2009. "Swedish Pirate Party Gets Election Boost." *Variety*, June 9, 25.

Michaelman, Frank I. 1996. Review of *Between Facts and Norms*, by Jürgen Habermas. *Journal of Philosophy* 93 (6): 307–315.

Miegel, Fredrik, and Tobias Olsson. 2008. From Pirates to Politicians: The Story of the Swedish File Sharers Who Became a Political Party. In *Democracy, Journalism, and Technology: New Developments in an Enlarged Europe*, ed. Nico Carpentier, Pille Pruulmann-Vengerfeldt, Kaarle Nordenstreng, Maren Hartmann, Peeter Vihalemm, Bart Cammaerts, Hannu Nieminen, and Tobias Olsson, 203–216. Tartu, Estonia: Tartu University Press.

Morozov, Evgeny. 2011. *The Net Delusion: The Dark Side of Internet Freedom*. New York: Public Affairs.

Mortazavi, Reza. 1997. The Right of Public Access in Sweden. *Annals of Tourism Research* 24 (3): 609–623.

Mosco, Vincent. 1989. *The Pay-Per Society: Computers and Communication in the Information Age*. New York: Ablex.

Mosco, Vincent. 2009. *The Political Economy of Communication*. Thousand Oaks, CA: Sage.

Mouffe, Chantal. 1999. Deliberative Democracy or Agonistic Pluralism? *Social Research* 66 (3): 745–758.

Moya, Jared. 2011. "UN Report: 3-Strikes Is a 'Violation of Human Rights.'" ZeroPaid, June 3. http://www.zeropaid.com/news/93617/un-report-3-strikes-is-a-violation-of-human-rights/ (accessed July 15, 2013).

MPAA (Motion Picture Association of America). 2010. "Theatrical Market Statistics 2010." http://mpaa.org/resources/93bbeb16-0e4d-4b7e-b085-3f41c459f9ac.pdf (accessed July 15, 2013).

Mueller, Milton L. 2004. *Ruling the Root: Internet Governance and the Taming of Cyberspace*. Cambridge, MA: MIT Press.

Mueller, Ursula. 1997. "A History of the Swedish Greens." *Synthesis/Regeneration* 13 (Spring). http://www.greens.org/s-r/13/13-08.html (accessed July 15, 2013).

Nitsche, Andreas. n.d. "LiquidFeedback Mission Statement." http://liquidfeedback.org/mission (accessed July 15, 2013).

NMPA (National Music Publishers' Association). 2001. NMPA International Survey of Music Publishing Revenues, 12th ed. Washington, DC. http://www.nmpa.org/media/surveys/twelvth/NMPA_International_Survey_12th_Edition.pdf (accessed July 15, 2013).

Norton, Quinn. 2006. "A Nation Divided over Piracy." *Wired*, August 17.

Nylander, Johan. 2011. "Internet Boosts Sweden's Economy 'Significantly.'" The Swedish Wire, May 20. http://www.swedishwire.com/economy/9934-internet-boosts-swedens-economy-significantly (accessed July 15, 2013).

O'Dwyer, Davin. 2009. "Swedish Swashbucklers." *Irish Times*, October 9.

OECD (Organization for Economic Cooperation and Development). 2008a. *Sweden*. OECD Economic Surveys. http://www.regeringen.se/content/1/c6/11/67/60/1844d532.pdf (accessed July 15, 2013).

OECD. 2008b. "Globalization in Services: From Measurement to Analysis." OECD Statistics Working Paper. http://dx.doi.org/10.1787/243156015316 (accessed July 15, 2013).

OECD. 2009. "The Impact of the Crisis on ICT and ICT-Related Employment." October. http://www.oecd.org/internet/ieconomy/43969700.pdf (accessed July 13, 2013).

Offe, Claus. 1985. New Social Movements: Challenging the Boundaries of Institutional Politics. *Social Research* 52 (4): 817–868.

Offe, Claus. 1990. Reflections on the Institutional Self-Transformation of Movement Politics: A Tentative Stage Model. In *Challenging the Political Order: New Social and Political Movements in Western Democracies*, ed. Russell J. Dalton and Manfred Kuechler, 232–250. New York: Oxford University Press.

Olsson, Tobias. 2008. The Practices of Internet Networking: A Resource for Alternative Political Movements. *Information Communication and Society* 11 (5): 659–674.

O'Neill, Michael. 2000. Preparing for Power: The German Greens and the Challenge of Party Politics. *Contemporary Politics* 6 (2): 165–184.

Open Rights Group. 2011. "IPRED." http://wiki.openrightsgroup.org/wiki/IPRED (accessed July 13, 2013).

PAC (Pierre Audoin Consultants). 2009. "D2—The European Software Industry: Economic and Social Impact of Software and Software-Based Services." http://cordis.europa.eu/fp7/ict/ssai/docs/20090730-d2-eu-ssbs-industry_en.pdf (accessed July 15, 2013).

Papenfuss, Mary. 2011. "Pirate Party Storms German Ship of State." Newser, September 20. http://www.newser.com/story/128956/pirate-party-storms-german-ship-of-state.html (accessed July 15, 2013).

Peñalver, Eduardo Moisés, and Sonia K. Katyal. 2010. *Property Outlaws: How Squatters, Pirates, and Protesters Improve the Law of Ownership*. New Haven, CT: Yale University Press.

Phillips, Leigh. 2009. "Pirates to Join Green or Liberal Groups in EU Parliament." EUObserver.com, June 3. http://euobserver.com/news/28237 (accessed July 13, 2013).

Pignal, Stanley. 2009. "Web Push Derails Europe Telecoms Reform." *Financial Times*, May 6.

Pirates de Catalunya. n.d. "Joint Complaint of Those Affected by the Closure of Megaupload Service." http://megaupload.pirata.cat (accessed July 15, 2013).

Pløger, John. 2010. Presence-Experiences: The Eventalisation of Urban Space. *Environment and Planning. D, Society & Space* 28 (5): 848–866.

PPCA (Pirate Party of Canada). 2011. "History." https://www.pirateparty .ca/about/history (accessed July 15, 2013).

PPI (Pirate Parties International). n.d. "About the PPI." http://www.pp -international.net/about (accessed July 15, 2013).

PPNZ (Pirate Party of New Zealand). 2011. "News." http://pirateparty .org.nz/news/ (accessed July 15, 2013).

PPS (Pirate Party of Sweden). 2012. "Pirate Party Principles." May 19. http://www.piratpartiet.se/politik/piratpartiets-principer/ (accessed July 15, 2013).

PP-UK (Pirate Party of the United Kingdom). 2012. "BT Blocks The Pirate Bay." June 20. http://www.pirateparty.org.uk/press/releases/2012/ jun/20/bt-blocks-pirate-bay (accessed July 15, 2013).

Puppis, Manuel. 2010. Media Governance: A New Concept for the Analysis of Media Policy and Regulation. *Communication, Culture & Critique* 3:134–149.

Putnam, Robert D. 1988. Diplomacy and Domestic Politics: The Logic of Two-Level Games. *International Organization* 42 (Summer): 427–460.

Quiggin, John. 2009. "Suicidally Strong IP?" *John Quiggin* (blog), June 10. http://johnquiggin.com/2009/06/10/suicidally-strong-ip/ (accessed July 15, 2013).

Raz, Guy. 2009. "Sweden's Pirate Party Fights Copyright Laws." *National Public Radio*, June 14. http://www.npr.org/templates/story/story.php ?storyId=105387231 (accessed July 15, 2013).

Reissman, Ole. 2009. "Pirates Plunder Germany's Big-Party Voters." Spiegel Online International, September 28. http://www.spiegel.de/ international/germany/election-raiding-party-pirates-plunder-germany -s-big-party-voters-a-651748.html (accessed July 15, 2013).

RIAA (Recording Industry Association of America). 2010. "2010 Year-End Shipment Statistics." http://76.74.24.142/548C3F4C-6B6D-F702-384C-D25E2AB93610.pdf (accessed July 13, 2013).

Robins, Kevin, and Frank Webster. 1999. *Times of the Technoculture: From the Information Society to the Virtual Life*. New York: Routledge.

Rodríguez, Clemencia, Dorothy Kidd, and Laura Stein, eds. 2009. *Making Our Media: Global Initiatives toward a Democratic Public Sphere*. Creskill, NJ: Hampton.

Rønning, Helge. 2003. Sweden, Denmark, Norway, Finland and Iceland, Status of Media In. In *Encyclopedia of International Media and Communications*, vol. 4, ed. Donald H. Johnston, 285–300. Boston, MA: Academic Press.

Rosen, Jeffrey. 2012. "The Right to Be Forgotten." *Stanford Law Review Online* 64:88–116. http://www.stanfordlawreview.org/online/privacy-paradox/right-to-be-forgotten (accessed July 15, 2013).

Rovira, Guiomar. 2009. *Zapatistas sin Fronteras: Las Redes de Solidaridad con Chiapas y el Altermundismo*. Mexico: Ediciones Era, S.A. de C.V.

Rusche, Herbert. 2009. "Demokratiesystem Deutschland—Ein Gespräch mit Herbert Rusche." Interview by Alexander Double. Piratig, September 15. http://piratig.de/2009/09/15/demokratiesystem-deutschland-ein-gesprach-mit-herbert-rusche (accessed July 15, 2013).

Russell, Adrienne. 2005. Myth and the Zapatista Movement: Exploring a Network Identity. *New Media & Society* 7 (4): 559–577.

Ryan, Johnny, and Caitriona Heinl. 2010. "Internet Access Controls: Three Strikes 'Graduated Response' Initiatives." Draft paper. Institute of International and European Affairs. http://www.iiea.com/documents/draft-overview-of-three-strikes-measures-nlm-study (accessed July 15, 2013).

Sassen, Saskia. 1999. Digital Networks and Power. In *Spaces of Culture: City, Nation, World*, ed. Mike Featherstone and Scott Lash, 49–63. Thousand Oaks, CA: Sage.

Schattschneider, E. E. 1988. *The Semi-Sovereign People: A Realist's View of Democracy in America*. Australia: Wadsworth Thomson Learning. First published by Holt, Rinehart and Winston, San Diego, CA, 1960.

Schiller, Dan. 2003. Digital Capitalism: A Status Report on the Corporate Commonwealth of Information. In *A Companion to Media Studies*, ed. Angharad N. Valdivia, 137–156. Malden, MA: Blackwell.

Schroeder, Stan. 2009a. "New EU Legislation Protects File Sharers—to a Certain Extent." Mashable, November 5.

Schroeder, Stan. 2009b. "The Pirate Bay sold for $7.8 million." Mashable, June 30.

Schroeder, Stan. 2010. "The Pirate Bay to RIAA: We Are Unsinkable." Mashable, May 18.

Schutz, Alfred, and Thomas Luckman. 1973. *The Structures of the Life-World*. Evanston, IL: Northwestern University Press.

Schweitzer, Eva. 2011. Normalization 2.0: A Longitudinal Analysis of German Online Campaigns in the National Elections, 2002–9. *European Journal of Communication* 26 (4): 310–327.

Sell, Susan. 2010. "The Global IP Upward Ratchet, Anti-Counterfeiting, and Piracy Enforcement Efforts: The State of Play." PIJIP (Program on Information Justice and Intellectual Property) Research Paper no. 15. American University Washington College of Law, Washington, DC. http://digitalcommons.wcl.american.edu/cgi/viewcontent.cgi?article=1016&context=research (accessed July 15, 2013).

Senate of France. 2009. "ACT No. 2009-669 of 12 June 2009 to Promote the Dissemination and Protection of Creation on the Internet." http://legifrance.gouv.fr (accessed July 15, 2013).

Shaw, Martin. 2005. Peace Activism and Western Wars: Social Movements in Mass-Mediated Global Politics. In *Global Activism, Global Media*, ed. Wilma DeJong, Martin Shaw, and Neil Stammers, 133–146. Ann Arbor, MI: Pluto Press.

Shaxson, Nicholas. 2012. *Treasure Islands: Uncovering the Damage of Off-shore Banking and Tax Havens*. New York: Palgrave Macmillan.

Shiffer, James. 1992. The Sacred Right to Wander in Sweden. *Scandinavian Review* 80 (1): 19–25.

Sitton, John F. 1998. Disembodied Capitalism: Habermas's Conception of the Economy. *Sociological Forum* 13 (1): 61–83.

Sjöberg, Ulrika. 1999. The Rise of the Electronic Individual: A Study of How Young Swedish Teenagers Use and Perceive the Internet. *Telematics and Informatics* 16 (3): 113–133.

Sjöström, Ulf. 2006. *Pirate Watch* (blog). http://piratewatch.blogspot.com/2006/06/pictures-from-piracy-demonstration.html (accessed July 15, 2013).

Smirke, Richard. 2011. "IFPI 2011 Report: Global Recorded Music Sales Fall 8.4%." *Billboard*, March 30.

Smith, Andy. 2007. European Commissioners and the Prospects of a European Public Sphere: Information, Representation, and Legitimacy. In *The European Union and the Public Sphere: A Communicative Space in the Making?*, ed. John Erik Fossum and Philip Schlesinger, 227–245. New York: Routledge.

Smith, Brett. 2010. "FSF Comment for the USTR's 2010 Special 301 Review." Free Software Foundation, February 17. http://www.fsf.org/licensing/2010-02-ustr-comment.html (accessed July 15, 2013).

Soares, Claire. 2009. "Sweden: We're Not Geeks Any More, Say Pirates of the Internet." *Independent*, June 9.

Socialhack. 2012. "Queer Geeks Panel at 28c3." *Hacking the Social Web* (blog), March 20. http://socialhack.eu/wp/2012/03/queer-geeks-panel-at-28c3 (accessed July 15, 2013).

Solomon, Lawrence. 2006. "Proportional Pirates; Proportional Representation Fosters Disproportional Theft." *Canada Financial Post*, October 6.

Spender, Lynn. 2009. "Digital Culture, Copyright Maximalism, and the Challenge to Copyright Law." http://handle.uws.edu.au:8081/1959.7/42615 (accessed July 15, 2013).

Statistisches Bundesamt. 2012. "Foreign Trade." Federal Statistical Office, June 22. https://www.destatis.de/EN/FactsFigures/NationalEconomy Environment/ForeignTrade/TradingPartners/Tables/OrderRankGermany TradingPartners.pdf?__blob=publicationFile (accessed July 15, 2013).

Strangelove, Michael. 2005. *The Empire of Mind: Digital Piracy and the Anti-capitalist Movement.* Toronto: University of Toronto Press.

Sundaram, Ravi. 2010. *Pirate Modernity: Delhi's Media Urbanism.* London: Routledge.

Sutton, Maira. 2011. "2011 in Review: Developments in ACTA." Electronic Frontier Foundation, December 27. https://www.eff.org/deeplinks/2011/12/2011-review-developments-acta#_ftnref2 (accessed July 15, 2013).

Swedish Environmental Protection Agency. n.d. http://www.swedishepa.se/ (accessed July 15, 2013).

Sweeny, Mark. 2012. "Ofcom Outlines New Anti-Piracy Rules." June 26. http://www.guardian.co.uk/technology/2012/jun/26/ofcom-outlines-anti-piracy-rules (accessed July 15, 2013).

Tarrow, Sidney. 1994. *Power in Movement: Social Movements, Collective Action, and Politics.* New York: Cambridge University Press.

Tarrow, Sidney. 2005. *The New Transnational Activism.* Cambridge: Cambridge University Press.

Tarrow, Sidney, and Donatella Della Porta. 2005. "Conclusion: Globalization, Complex Internationalism, and Transnational Contention." In *Transnational Protest and Global Activism,* ed. Donatella Della Porta and Sidney Tarrow, 227–246. Lanham, MD: Rowman and Littlefield.

TASS (ITAR-TASS World Service). 2009. "Radical Forces in EU Parliament—Dangerous Trend." June 8.

Taylor, Simon. 2009. "Swedish Pirate Party MEP Joins the Green Group." *European Voice,* June 25. http://www.europeanvoice.com/article/2009/06/swedish-pirate-party-mep-joins-the-green-group-/65332.aspx (accessed July 15, 2013).

Terranova, Tiziana. 2001. Demonstrating the Globe: Virtual Action in the Network Society. In *Virtual Globalization: Virtual Spaces/Tourist Spaces*, ed. David Holmes, 95–113. New York: Routledge.

Thornburgh, Nathan. 2012. "Christiania: The Free Town That Is about to Be Sold." June 28. http://world.time.com/2012/06/28/christiania-the-free-town-that-is-about-to-be-sold/ (accessed July 15, 2013).

Tilly, Charles. 2006. *Regimes and Repertoires*. Chicago: University of Chicago Press.

TorrentFreak. 2009. "Pirate Party Gets Second Seat in European Parliament." November 4. http://torrentfreak.com/pirate-party-gets-second-seat-in-european-parliament-091104 (accessed July 15, 2013).

TorrentFreak. 2010a. "The Pirate Party Becomes the Pirate Bay's New Host." May 18. http://torrentfreak.com/the-pirate-party-becomes-the-pirate-bays-new-host-100518/?utm_source=feedburner&utm_medium=feed&utm_campaign=Feed%3ATorrentfreak%28Torrentfreak%29(accessed July 15, 2013).

TorrentFreak. 2010b. "Wikileaks Cable Shows US Involvement in Swedish Anti-piracy Efforts." December 8. http://torrentfreak.com/wikileaks-cable-shows-us-involvement-in-swedish-anti-piracy-efforts-101207/?utm_source=feedburner&utm_medium=feed&utm_campaign=Feed%3A Torrentfreak %28Torrentfreak%29 (accessed July 15, 2013).

TorrentFreak. 2012a. "MegaUpload Users Plan to Sue the FBI over Lost Files." January 26. http://torrentfreak.com/megaupload-users-plan-to-sue-the-fbi-over-lost-files-120126 (accessed July 15, 2013).

TorrentFreak. 2012b. "Massive Street Protests Wage War on ACTA Anti-Piracy Treaty." February 11. http://torrentfreak.com/massive-street-protests-wage-war-on-acta-anti-piracy-treaty-120211 (accessed July 15, 2013).

Touloumis, Tara. 2009. Buccaneers and Bucks from the Internet: Pirate Bay and the Entertainment Industry. *Seton Hall Journal of Sports and Entertainment Law Review* 19 (1): 253–266.

Touraine, Alain. 1988. *Return of the Actor: Social Theory in Postindustrial Society*. Minneapolis: University of Minnesota Press.

TPB (The Pirate Bay). 2003. *Histoire (Paris)*. http://thepiratebay.sx/history/2003 (accessed July 13, 2013).

Tuccille, J. D. 2009. "Pirates Prepare to Hoist Jolly Roger in Germany." *Disloyal Opposition* (blog), September 29. http://www.tuccille.com/blog/2009/09/pirates-prepare-to-hoist-jolly-roger-in.html (accessed July 13, 2013).

Turner, Fred. 2006. *From Counterculture to Cyberculture: Stewart Brand, the Whole Earth Network, and the Rise of Digital Utopianism*. Chicago: University of Chicago Press.

UPI. 2009. "Pirate Bay Sentences Prompt Protests." April 19.

U.S. Department of State. 2009b. "Special 301 for Sweden: Post Recommendation." Stockholm Embassy Cable 000141. http://www.cabledrum.net/diff/09STOCKHOLM141 (accessed July 15, 2013).

Vaidhyanathan, Siva. 2005. Celestial Jukebox: The Paradox of Intellectual Property. *American Scholar* 74 (2): 131–135.

Vandystadt, Nathalie. 2009a. "Council Set to Reject Amendment 138 as EP Names Negotiators." *Europolitics*, September 25. http://www.europolitics.info/council-set-to-reject-amendment-138-as-ep-names-negotiators-art248979-9.html (accessed July 15, 2013).

Vandystadt, Nathalie. 2009b. "Amendment 138: EU Wants to Avoid Confrontation." *Europolitics*, November 4. http://www.europolitics.info/amendment-138-eu-wants-to-avoid-confrontation-art253326-1.html (accessed July 15, 2013).

Varnelis, Kazys, ed. 2008. *Networked Publics*. Cambridge, MA: MIT Press.

Vichot, Rhea. 2012. "'We Do Not Forgive. We Do Not Forget': An Anonymous Ethos of Lulz and Epideictic Rhetoric." Paper presented at the International Communication Association conference, Phoenix, AZ, May.

Vihriälä, Vesa. 1991. Spotlight on Sweden. *OECD Observer*, no. 168: 43–44.

Weibull, Lennart. 2003. The Press Subsidy System in Sweden: A Critical Approach. In *Contesting Media Power: Alternative Media in a Networked World*, ed. Nick Couldry and James Curran, 89–107. Lanham, MD: Rowman and Littlefield.

Willis, Andrew. 2009. "Brussels Claims Failed Business Model Is Causing Online Piracy." EUObserver.com, July 9. http://euobserver.com/creative/28438 (accessed July 13, 2013).

Wirtén, Eva Hemmungs. 2006. Out of Sight and Out of Mind: On the Cultural Hegemony of Intellectual Property (Critique). *Cultural Studies* 20 (2–3): 282–291.

Woldt, Marco. 2009. "Pro-piracy Parties Gain Foothold across Europe." CNN.com, July 22. http://edition.cnn.com/2009/SHOWBIZ/Movies/07/22/pirate.party.christian.engstrom/index.html (accessed July 15, 2013).

WordIQ. 2010. "List of Swedish Government Owned Companies." http://www.wordiq.com/definition/List_of_Swedish_government_owned_companies (accessed July 15, 2013).

Wyrembek, Christian. 2011. "Heute Basis, morgen Parlament." Taz, September 14. http://www.taz.de/Wahlkampf-der-Piratenpartei/!77894 (accessed July 15, 2013).

Yar, Majid. 2008. The Rhetorics and Myths of Anti-piracy Campaigns: Criminalization, Moral Pedagogy, and Capitalist Property Relations in the Classroom. *New Media & Society* 10 (4): 605–623.

Zhao, Shanyang. 2004. Consociated Contemporaries as an Emergent Realm of the Lifeworld: Extending Schutz's Phenomenological Analysis to Cyberspace. *Human Studies* 27:91–105.

Zittrain, Jonathan. 2008. *The Future of the Internet and How to Stop It.* New Haven, CT: Yale University Press.

Index

Note: t after a page number indicates a table.